A Pen and Paper Introduction to Statistics

Statistics is central in the biosciences, social sciences and other disciplines, yet many students often struggle to learn how to perform statistical tests, and to understand how and why statistical tests work. Although there are many approaches to teaching statistics, a common framework exists between them: starting with probability and distributions, then sampling from distribution and descriptive statistics and later introducing both simple and complex statistical tests, typically ending with regression analysis (linear models).

This book proposes to reverse the way statistics is taught, by starting with the introduction of linear models. Today, many statisticians know that the one unifying principle of statistical tests is that most of them are instances of linear models. This teaching method has two advantages: all statistical tests in a course can be presented under the same unifying framework, simplifying things; second, linear models can be expressed as lines over squared paper, replacing any equation with a drawing.

This book explains how and why statistics works without using a single equation, just lines and squares over grid paper. The reader will have the opportunity to work through the examples and compute sums of squares by just drawing and counting, and finally evaluating whether observed differences are statistically significant by using the tables provided. Intended for students, scientists and those with little prior knowledge of statistics, this book is for all with simple and clear examples, computations and drawings helping the reader to not only do statistical tests but also understand statistics.

Antonio Marco, Ph.D., is Senior Lecturer in the School of Life Sciences at The University of Essex. Dr Marco obtained a degree in Biological Sciences and a doctorate in Genetics from the University of Valencia and has a Graduate Diploma in Mathematics from the University of London (LSE). He has been a lecturer at the University of Essex since 2013, teaching statistics, mathematics, genomics and bioinformatics, and is the course director of the BSc Genetics and the MSc Health Genomics programmes.

A Pen and Paper Introduction to Statistics

Antonio Marco

CRC Press is an imprint of the
Taylor & Francis Group, an **informa** business
A CHAPMAN & HALL BOOK

Designed cover image: © Antonio Marco

First edition published 2024
by CRC Press
2385 NW Executive Center Drive, Suite 320, Boca Raton FL 33431

and by CRC Press
4 Park Square, Milton Park, Abingdon, Oxon, OX14 4RN

CRC Press is an imprint of Taylor & Francis Group, LLC

ISBN: 978-1-032-50511-4 (hbk)
ISBN: 978-1-032-50510-7 (pbk)
ISBN: 978-1-003-39882-0 (ebk)

DOI: 10.1201/9781003398820

Typeset in Palatino
by Apex CoVantage, LLC

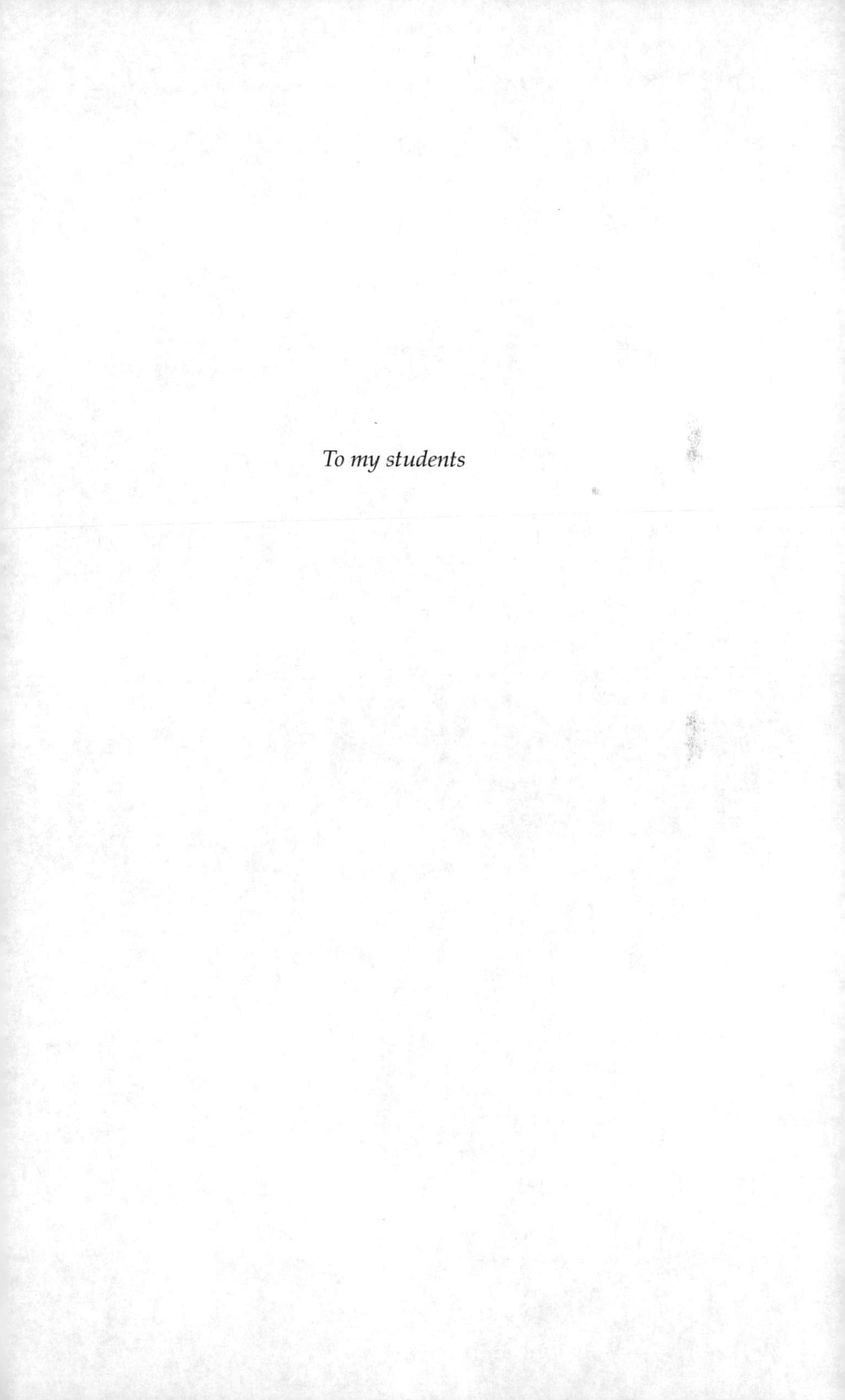

To my students

Contents

Acknowledgements

Like many undergrads, I started my studies without any interest in statistics. As a matter of fact, I failed statistics in my first year. Five years later, I was developing my own statistical methods, and ten years after I failed first-year statistics, I was teaching (and preaching) about statistics good practice. Now, when I teach statistics to first-year students, I feel that many of them have the same interest in statistics as I had when I was a student: none. Yet, I am aware that their success as students and professional scientists will heavily depend on how well they develop their quantitative and statistical skills. This has been driving me to constantly adapt my teaching methods to ensure that students engage in the teaching events without losing any rigour in the statistical methods they learn.

The first time I thought about teaching statistics using pen and paper, rather than formulas, was during the Essex Summer School in 2015, while teaching linear models. However, it wasn't until a conversation I had with Uli Bechtold and Jordi Paps that I thought on writing down this approach as a proper textbook. During that conversation, I was proposing to adapt an existing method to perform statistical tests over evolutionary trees, drawing lines and squares over graph paper during my explanations. Uli and Jordi unanimously sentenced: 'you should write a book about this'. Well, here it is.

Thus, my first acknowledgement is to Uli and Jordi for listening to my statistics speeches for years and for supporting the crazy idea of writing a book about it. I am also very grateful to Ben Skinner, who critically read the first draft chapters of this book, and with whom I have spent countless hours discussing best practices to teach data analysis. Thanks also to the organisers of the R course at the Essex Summer School (and its variations) through the years for inviting me to teach statistics with total freedom, including the use of pen and paper: Leo Schalkwyk, Radu Zabet, Dave Clark and Ben Skinner. I'm also grateful to all my academic colleagues for their continuous support. I would also like to thank the Taylor and Francis editorial team, particularly to Lara Spieker, for their support and for making the review and production process smooth and effortless. Last (and certainly not least), a big thanks to all my students, those who showed interest, and those who didn't, as all have contributed, to one or another degree, to the development of this work.

A significant part of this book was written during a study leave (sabbatical) granted by the University of Essex during the academic year 2021–2022.

Colchester, September 2023

Before We Start . . .

We are going to do a small exercise to better understand the approach taken in this book. The aim is to present statistics to a certain level of complexity without the use of a single equation. Instead, we will replace equations and formulas with lines and squares drawn on graph paper. So, get a graph paper notebook, a ruler and a pencil, and let's do some statistics.

FIGURE 0.1
An unusual yet powerful statistics 'software'. All you will need to learn statistical analysis is graph paper, a ruler and a pencil (or several colour pencils for the full experience). No computers, no calculators and no formulas here.

As a keen observer, you noticed that old people sleep less time than young people, and you are determined to provide convincing evidence for that. However, you can only have access to the sleeping times of a few of your close friends. You observe that two young people sleep seven and nine hours, respectively, every night, whilst the two oldest sleep five and seven hours each. Are these four observations enough to affirm that older people sleep less hours than young people? Let's work it out.

In a graph paper, draw a vertical line and label each square unit as one hour of sleep from one to ten. Add also a horizontal line at the bottom, as in the following cartoon:

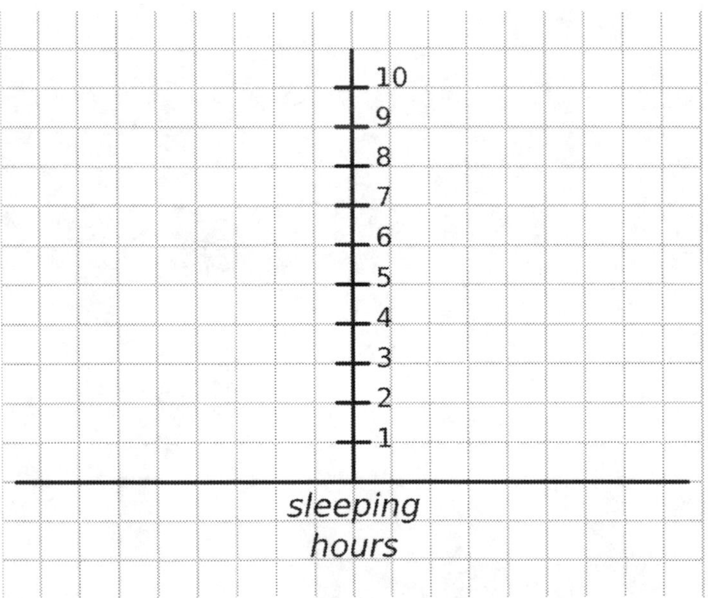

Now, draw on the left two points representing the hours of sleep of the young people and on the right the hours of sleep of the old people:

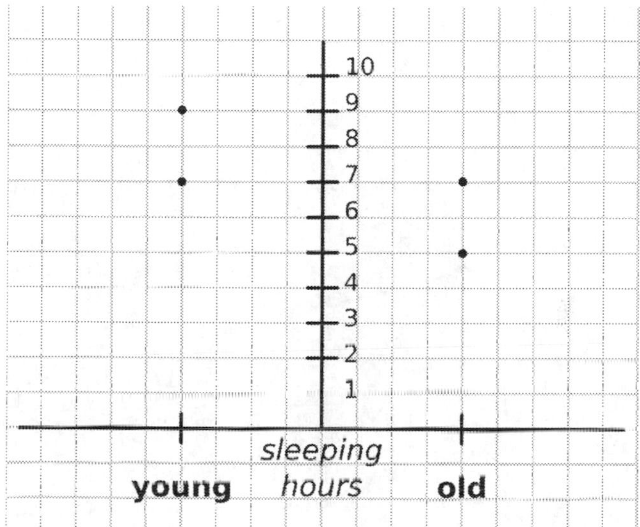

Place a small dot between each of the two sleep times right in the middle and draw a connecting line. These small dots are the estimation of the average sleeping times in each group:

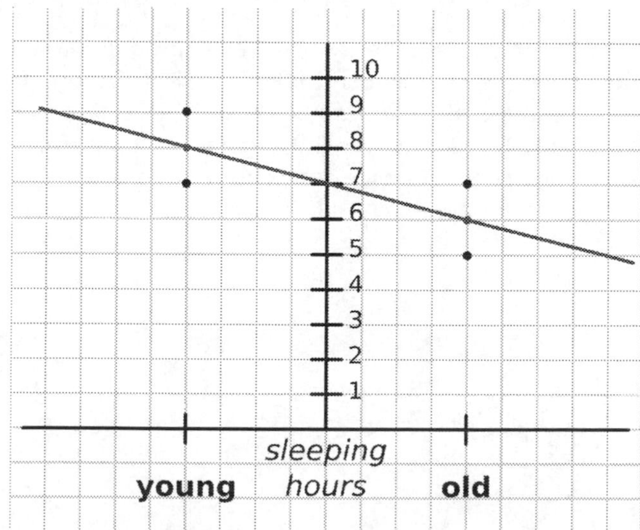

Now draw and colour for each observation one square connecting the observation and the line:

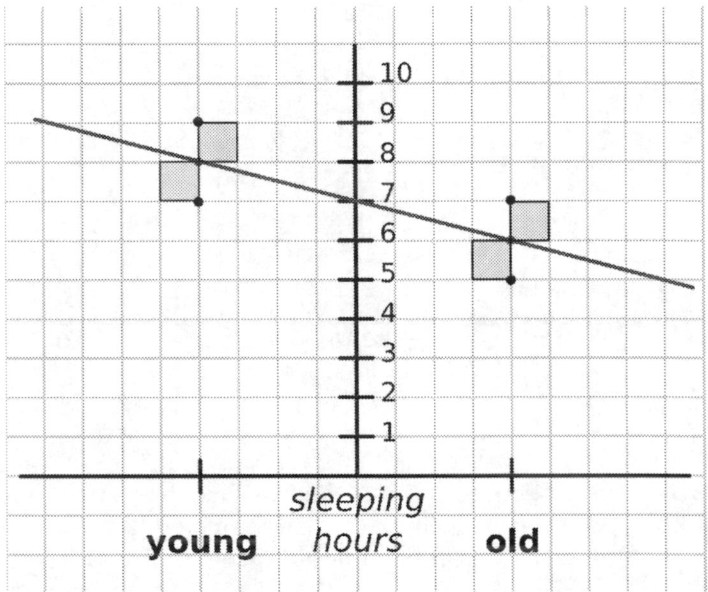

How many little (grid) squares have you coloured in total? This is called the sum of squares which in this case is 4. Draw the same graph again, but this time add a horizontal line cutting the ruled vertical line at 7 (this represents the average sleeping time across all individuals). Then, connect each observation to the horizontal line with a square. Notice that, in this case, each drawn and coloured square will be made of smaller grid squares. Count the grid squares – you will find out that in this case, the sum of squares is exactly 8:

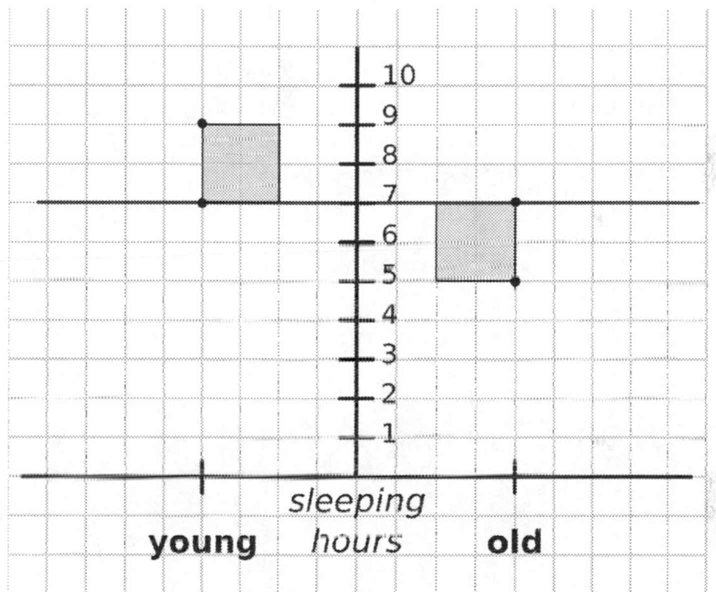

From the plots you just generated you can conclude that, on average, older people sleep less than young people (this is visually represented by the diagonal line). But how real is this difference? Could it be that the observed difference is just due to random sampling? In this book, I will explain that by comparing the number of squares of the two graphs, you can quantify how wrong (or right) you are. In this particular case, the sums of squares reveal that if you claim that old people sleep less than young people, you have a 29% chance of being wrong. How this figure is computed is the topic of this book, and you will learn how to find it out.

This example illustrates what statistics is about. First you estimated the average sleeping times from a limited sample (the horizontal line) and for each group to be compared (the connected small dots). Then you quantified precisely what are the chances of this difference to be the product of random sampling. If you follow the examples in this book, by Lesson 5, you will understand how we computed this percentage. And as I promised, you won't have to learn a single mathematical equation in the process. Ready for some more statistics? Off we go!

1

What Is Statistics?

How do we know things? For instance, how do we know that the Earth travels around the Sun? Or that smoking causes cancer? Or that hares are faster than tortoises? One could argue that this is easily observable, but actually we don't see the Earth travelling around the Sun yet we know it does. On the other hand, one can claim that this is just common sense. But unfortunately, 'common' sense is heavily subjective. How do we know things, then?

This seemingly trivial question has puzzled philosophers for millennia. Just for the record, the branch of philosophy that deals with the origin of knowledge is called epistemology, but many texts refer to it as simply 'theory of knowledge'. In the Western tradition, we roughly identify two schools of thought: rationalism and empiricism. *Rationalists* believed that knowledge is rational: reason is the source of knowledge. That is, if you think hard enough, you will find the eternal truth that explains everything. Importantly, you should not trust your senses. The classic example is that of René Descartes, who, after years of deep thinking and of avoiding any influence by any of his senses, concluded that he existed (that's genius!).

However, strict rationalism and neglecting observations are at odds with modern scientific thinking. Bertrand Russell once wrote:

> Observation versus Authority: To modern educated people, it seems obvious that matters of fact are to be ascertained by observation, not by consulting ancient authorities. But this is an entirely modern conception, which hardly existed before the seventeenth century. Aristotle maintained that women have fewer teeth than men; although he was twice married, it never occurred to him to verify this statement by examining his wives' mouths.
>
> (B. Russell *The Impact of Science in Society* 1952)

In part responding to the issues of rationalism, a new school of thought emerged, the so-called *empiricism*. For empiricists, seeing

DOI: 10.1201/9781003398820-1

is believing. Experiments are central to science and you can only obtain knowledge by observing and testing. In other words, if you want to understand the world, you have to get your hands dirty. Indeed, science is empirical.

Statistics

In science, we need to observe and quantify things, and then use our observations to extract knowledge. That is precisely why you need statistics. So, what's statistics? A definition will be handy here, but it's not easy to find. You will probably find as many definitions as people you ask and, in most statistics textbooks, you won't even find a definition. But I can't leave you without a definition in a book that aims to introduce statistics in an intuitive way! So, before that, let me start with an example.

Let's suppose you want to know whether hares are, on average, faster than tortoises. The answer seems obvious, but in reality, you will never be totally sure as you will need to measure how fast are all existing hares and tortoises, an impossible task. Instead, you take measurements from the two hares you captured in the field and the three tortoises from the local pet store. Are five observations enough to conclude that hares are faster than tortoises? Certainly, it'll be a risky statement. Would 1,000 hares and 1,000 tortoises be enough? How confident would you be about your claims? Are the speeds of your selected individuals representative of the actual speeds of the whole set of hares and tortoises in the world. If not, how accurate is the estimation?

All these questions can be answered using statistical methods. That is, statistics is a discipline that deals with collected data from observations and/or experiments (here's the connection with empiricism) to infer characteristics of the whole population (i.e. to infer knowledge). But we cannot be 100% sure of any inferred knowledge as we extracted it from a subset. Fortunately, statistical methods allow us to quantify the degree of confidence in our results. With statistics, you will be able to declare, for instance, that 'hares are faster than tortoises with a 99.97% confidence'. In simple words, *statistics is the science of uncertainty*.

Measurements

If we want to do any statistics at all, we first need to measure things (after all, science is empirical). In technical terms (and I'm afraid we have to define a number of them in this lesson), a collection of measurements of a specific characteristic is called a *variable*. For instance, the heights of a group of people, the measured speeds of a set of hares and the number of coins in the pockets of a bunch of kids are examples of variables.

Variables are of different types. A *discrete* variable is a variable that can only take fixed values (usually integers) that are sorted. For instance, the number of coins in a pocket. One can have 0, 1, 2, 3 or more coins, but no intermediate values (we assume here that half coins are not an option, although at some point in history halves and quarters of coins circulated frequently). Also, 1 is smaller than 2, which is smaller than 3 and so forth. The analysis of discrete variables is the subject of Lesson 13. A similar type of variables are those called *ranked*, which take fixed values, are sorted, but the exact values are not relevant. For instance, if we sort a group of 20 mice by size and give them a number from the biggest (1) to the smallest (20), in this case, sizes are ranked, as 1 is bigger than 2, which is bigger than 3 and so forth. However, the difference between 1 and 2 needs not to be the same as the difference between 2 or 3. We will be working with ranked variables in Lesson 12, in the context of non-parametric statistics. Often, we are interested in variables that are discrete but not sorted. These variables are called *categorical* (as in different categories), like for instance eye colour, nationality or marital status.

The most common type of variable you will find in statistics is a *continuous* variable, which is defined as variables for which no matter how close two distinct values are, there can always be a value in the middle. For instance, weight if one individual is 67 kilos 457 grams and another is 67 kilos 458 grams, there always exists the possibility that someone weighs something in the middle (like 67 kilos 457 grams 856 milligrams). In plain terms, anything that is not discrete, ranked or categorical is generally a continuous variable. We will work with continuous variables in most of the examples developed in this book (specifically from Lesson 6 till Lesson 12).

This classification of variables is quite handy to find out which sort of statistical analysis we can apply to our variables.

Samples and Populations

As we already discussed, we cannot measure the average speed for all hares and tortoises in the world. Likewise, we cannot test a drug in all ill patients in the world and then also in the rest of the healthy human population as a control. We just can't! For that reason, precisely we need statistics, to be able to infer what the whole looks like by analysing only a part of it. In this context, we need to introduce two important definitions: populations and samples. A *population* is a group of things that we are interested in, usually a measure of something. For instance, the height of all adult British males. A *sample* is a set of the population that we have collected and measured, for instance, the height of 1,000 adult British males randomly selected from the streets.

In statistics, we are interested in estimating how the population looks like by taking a sample. This is the realm of the so-called *statistical inference*. For instance, if we want to know the average height of adult British males (population), we won't be able to compute it unless we measure all adult British males. But we can measure a representative subset, the sample (let's say 1,000 individuals), and estimate the population average height using the sample average height. As we shall see, estimates computed from a sample are not always the best estimates for the whole population, but we have ways to correct them.

Finally, we want to be able to support a scientific statement from our samples that will apply to the whole population. For instance, is this drug good for curing cancer? As we cannot try the drug in all cancer and non-cancer patients, we may limit our analysis to 100 individuals with cancer and 100 controls, plus an equal number of patients where a *placebo* treatment is given. How this is done will be covered throughout the book, and it is what we call *statistical hypothesis testing*.

A First Look at the Data

After we collect our samples and before we do any statistical analysis, the first step is usually to look at the data. Obviously, a large list of numbers is not particularly intuitive. But there are many graphical approaches that make data interpretation easy.

The first thing we want to know is how our data are distributed. That is, how frequent is each value in our sample. To do so, we often plot a *histogram*. The recipe is pretty simple: sort your sample measurements and count how many times you find every value. Then, on a graph paper, draw a horizontal line labelling all sample measurements and draw a bar representing the frequency of each value. For instance, we measured the speed of 15 hares with these values in miles per hour: 9, 12, 14, 16, 16, 18, 20, 22, 22, 23, 25, 28, 28, 32 and 38. The histogram will look like that:

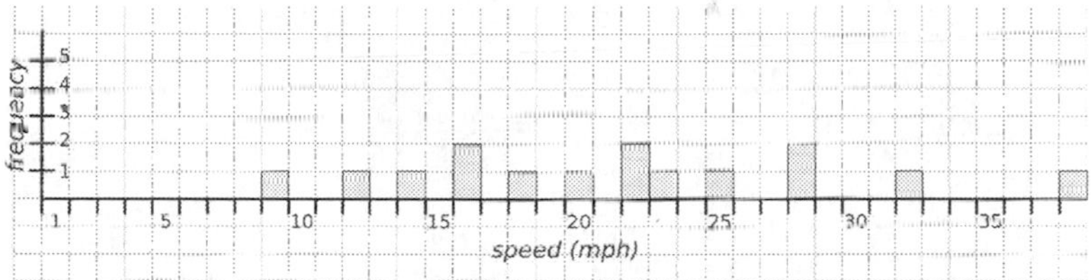

The problem is that, if we work with a limited number of data points or with continuous variables, the frequency of each value will be most likely one (as in the previous plot). In these cases, we define something called the *bin size*, which is a fixed range in which we group our values. For instance, if we plot the speeds of hares using a bin size of 5, the plot will look like that:

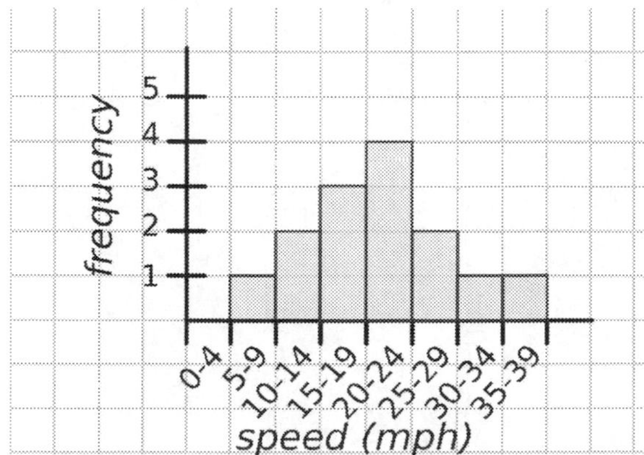

The shape of the histogram resembles a mountain, or a bell, and that's actually characteristic of most variables in nature.

Statisticians call this type of distribution a *normal distribution*. This concept is central to statistics, and we will be discussing it several times throughout the book.

Another way of looking at the distribution of a sample is by the use of a *boxplot*. In a boxplot, we represent the most 'central' value of a sample and how spread the data is. The advantage of this sort of plot is that it allows you to compare multiple samples in the same graph. For a sample, you can build a boxplot following a simple recipe, which we will be using with the previous sample of hares' speeds.

1) Sort the values numerically.

   9-12-14-16-16-18-20-22-22-23-25-28-32-32-38
2) Find the value in the middle.

   9-12-14-16-16-18-20-*22*-22-23-25-28-32-32-38
3) Find the values between the first and middle values, and between the middle and last values.

   9-12-14-*16*-16-18-20-22-22-23-25-*28*-32-32-38
4) On a graph paper, draw a box whose height goes from the smallest to the largest of the two values you found in the previous step (these are technically called the quartiles, as we shall see in the next lesson).

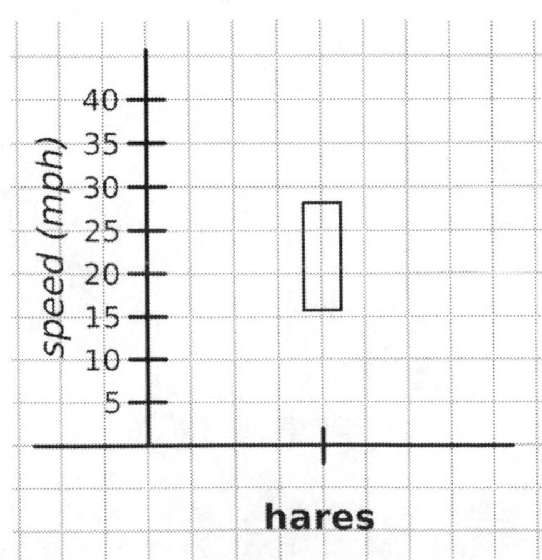

5) Draw a horizontal line inside the box where the middle value from step 2 is.

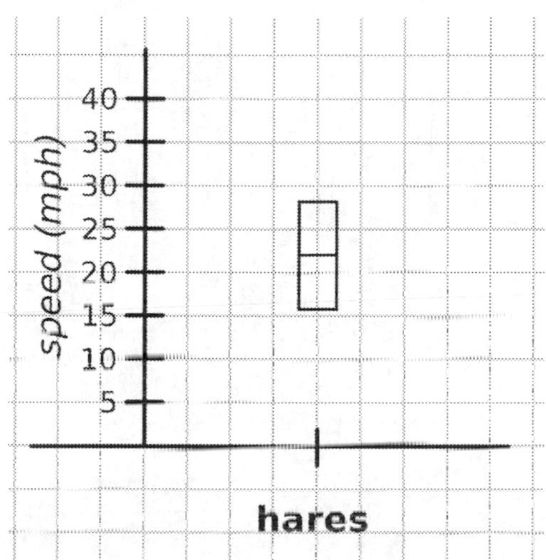

6) Draw two small horizontal lines, one representing the smallest value and the other representing the largest value. Join them to the main box and, voilà, you built your first boxplot:

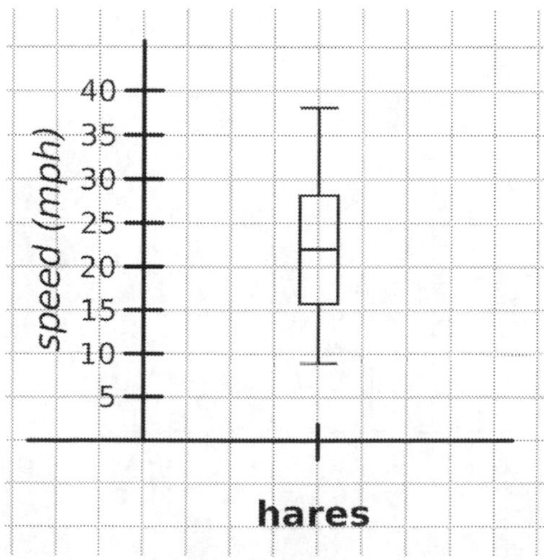

The usefulness of the plot is evident when you compare two or more samples. For instance, the boxplot below represents measured speeds from a sample of hares and a sample of tortoises:

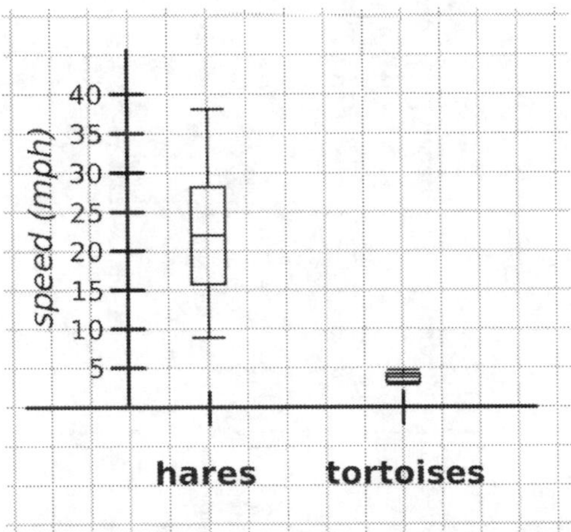

A similar type of plot but slightly different is the *barplot*. These plots have bars whose height indicates the average value of the sample. Often, they have some lines indicating how spread the data is (we will be working out some of these values in Lesson 2). For instance, we can build a barplot using the hares and tortoises' speeds as before:

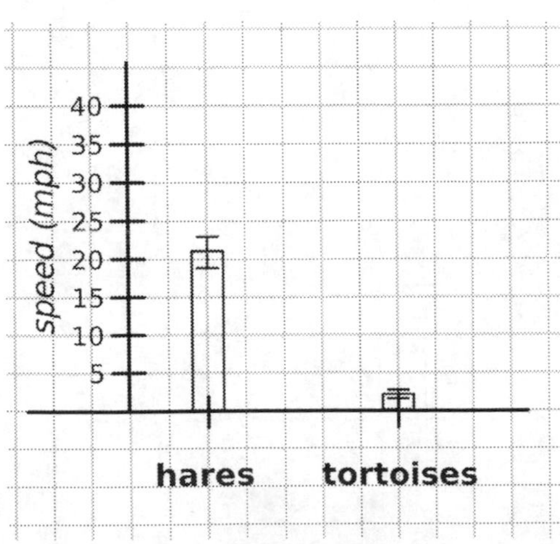

Barplots are being used less and less in favour of boxplots, so we won't be spending more time on them.

The previous plots tell us about what the data look like, but we also want to find relationships between different samples. A *scatterplot* compares the values of two paired samples. By paired, we mean that each value in one sample has a counterpart in the other sample. For instance, you measure the speeds of five hares and five tortoises in five consecutive races, in which one hare and one tortoise compete with each other. The pairs of speeds in miles per hour are 15 | 0.3, 30 | 0.4, 25 | 0.5, 30 | 0.7 and 30 | 0.8. A scatterplot will look like this (notice that the value of each square is different between the vertical and horizontal axes):

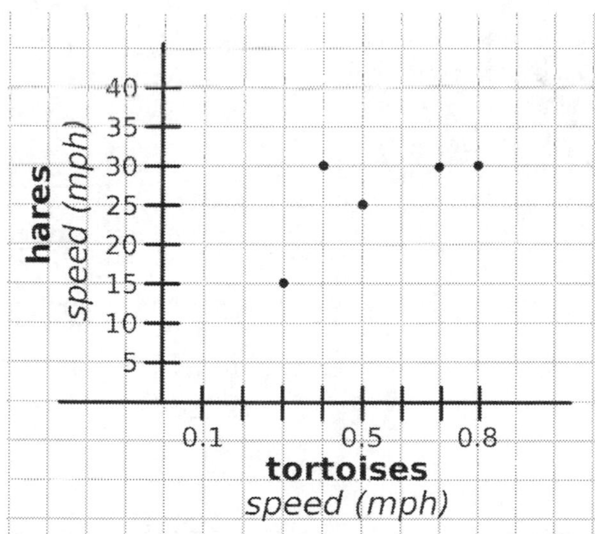

We will be plotting a lot of scatter plots in this book. Indeed, you already built one in the introductory exercise.

Last, but not least, a useful piece of advice: never, ever, ever use *pie charts*. Pie charts are the evil instrument that politicians and business people use to cheat on us, to make us believe that a product or a political party is better than that of the competitors. If you want to be honest, stay away from pie charts.

TRY ON YOUR OWN

Alice never cleans up after her and, after months of accumulating food leftovers in the kitchen, she has now an infestation of mice at home. Bob also has mice, but in his case, he just likes animals and keeps a few mice at home. Bob was horrified when he saw Alice's mice, not because of the mice themselves, but because of their diet. 'Alice', said Bob, 'you can't keep your mice in such an unhealthy condition, eating pizza leftovers all the time. Give them some proper healthy food, they are getting really fat'. 'No, they are not fat, they are as slim as yours', replied Alice. Who's right? Let's have a look at the mice weights.

Bob keeps an updated record of his nine mice's weights, which in grams are: 13, 21, 11, 20, 22, 24, 16, 18 and 17. Alice weighed the seven mice, she found in the kitchen scale (hygiene is not a concern to her) giving the following values in grams: 29, 28, 13, 15, 17, 28 and 17. On a graph paper do the following:

- Draw a histogram for each group of mice in the same graph (using a different colour for each sample, and a bin size of 5).
- Build a boxplot comparing the two mice samples.

By looking at the graphs, what can you tell about the weights of Alice's and Bob's mice?

Next

So, are hares faster than tortoises? Or, are Alice's mice fatter than Bob's mice? From the graphs, it looks like both responses are affirmative. But how confident are we in each case? To do so, we needed to quantify our observations beyond graphs, and that's what we will be doing in the next lesson.

2

Know Your Samples

Time to take some measurements. As we discussed in the previous lesson, we only take measurements from a sample of the whole population. In this lesson, we will discuss how we describe samples.

Summary Statistics

Here I have to introduce another technical definition, the concept of a *statistic*. But this time, I'm not talking about statistics as a science, but a statistic as a number. A *summary statistic* is a function that takes multiple measurements from a sample and produces a single number (at least in the cases we are covering here); that is, it summarizes a large collection of values in a single value. For instance, the well-known average value is a single number computed from a larger sample that summarizes one of the characteristics of the sample.

The first type of summary statistics to be considered are those called summary statistics of *centrality*. These statistics tell us about the general tendency of a sample. The most used is the *mean* or *average* (we will be using both terms interchangeably), which is the sum of all observations divided by the number of observations. If I have four pounds in my pocket and you have eight, we have on average six pounds. But the mean is not always the best way to summarize the tendency of a sample. For instance, if both Alice and Bob have four pounds each and Carol has 70, we can say that, on average, each person has 26 (you equally split the total of 78 pounds among three people), but you will agree that this number can lead to confusion. A more meaningful statistic when extreme values exist is the *median*, which sorts all values from the sample and takes the one in the middle (4-4-70), so the

DOI: 10.1201/9781003398820-2

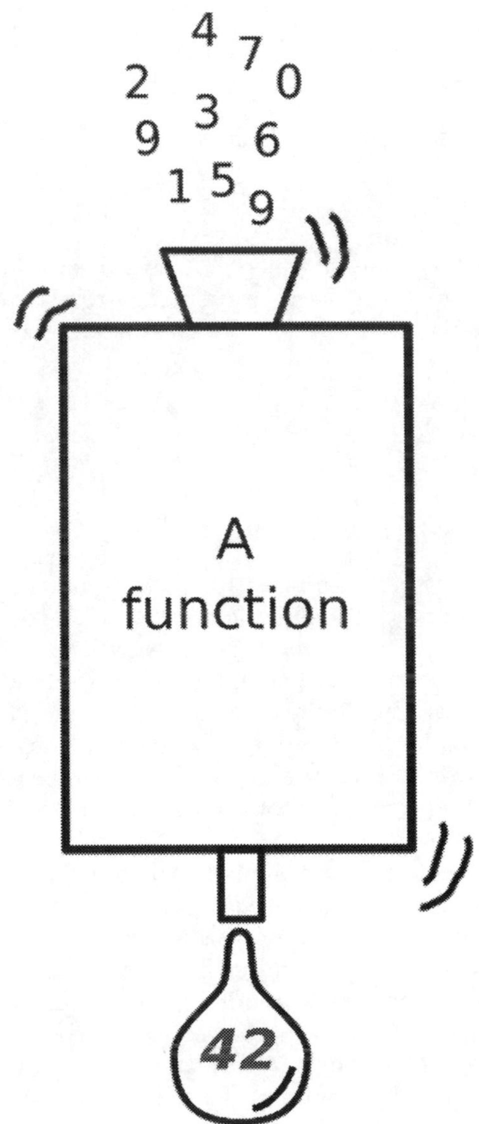

FIGURE 2.1
Cartoon representation of a function. A mathematical function is a formula that takes several numbers, makes some computations and returns an output. In the cases we cover in this book the output is a single number. However, in more advanced setups, some functions produce lists of numbers, or even more complex objects like other formulas.

median number of coins will be 4. A third measure of centrality (yet not used that much) is the *mode*, which is just the most common value (in this example, the mode will be again 4).

But summary statistics of centrality are often not enough. Let's imagine that Alice and her four sisters have a weekly allowance of four pounds each. Bob and his four brothers have, instead, a different weekly allowance according to their ages: 2, 3, 4, 5 and 6 pounds, respectively. Alice and her siblings make on average four pounds a week, so do Bob and siblings. However, there is some variation among Bob and his siblings but not among Alice's relatives. To quantify these differences, we use *statistics of dispersion*.

One simple way of measuring dispersion is to find out the *range* of values, that is, the difference between the greatest (maximum) and the smallest (minimum) values. So Alice and siblings' have a range allowance of 0 pounds, and Bob and siblings' have a range allowance of four pounds (the difference between six and two pounds). But the range is not very informative. In the previous section, I told you about the median, that is, the value right in the middle. Often, we also report the value in the middle between the smallest value and the median, and the value in the middle between the median and the greatest value. We call these two numbers the *first quartile* and the *third quartile*, respectively. Again, these values are not very useful to do statistics, but are quite handy for data visualization, as we did in the previous lesson. (When we plotted the boxplot, the box was drawn between the two quartiles, the line in the middle was the median, and the two additional segments indicated the range.)

There's a statistic of dispersion that better captures the variability of the data, and that we will use in this book, this is a statistic called *variance*, but we will need a whole section only to introduce this important concept.

Variance: The Mean Sum of Squares

Let's start having a look at Bob and his siblings' allowances. The mean value is 4 but, how far from the mean are each of the salaries? We will ignore whether the difference is positive or negative. Thus, the difference between the mean salary and the actual salary for the five members of the family will be (in pounds): 2, 1, 0, 1 and 2. A simple measure of dispersion can be the mean difference, in this case, one pound and 20 pence. But we rarely use it (if ever). Instead, we are going to focus on squares, as in the rest of this book.

The *variance* is defined as the mean of the *sum of squares*. There's a mathematical formula for that, but I promised that I won't use any formula so, let's work out the variance as a bunch of actual squares. If we plot a line representing the mean allowance and dots for the individual allowances, we can find the distance between each amount and the average graphically, as we computed before. Instead, let's complete the squares for each one of the segments connecting the actual measurements and the mean value, as in the graph below:

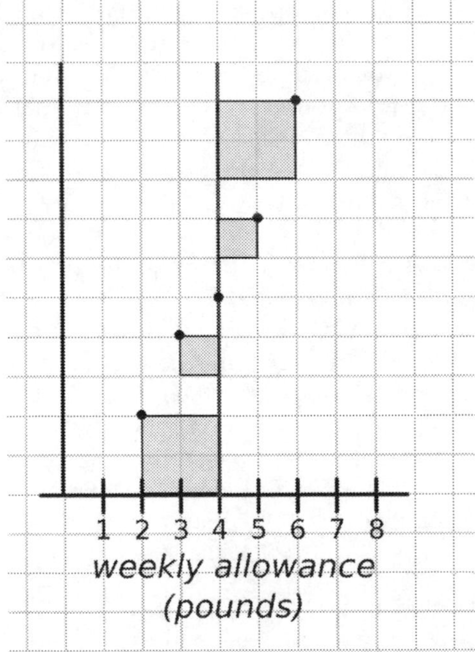

You can easily compute the area of each square as the number of little squares on the graph paper within each drawn square, and then the mean area of the squares, which is exactly the variance. In this example, the sum of squares is 10 and therefore the variance is 2.

Why squares and not cubes? Or hypercubes? Or just lines? The main reason is a property of the sum of squares called *additivity*, but you don't need to worry at this moment as we will devote the whole of Lesson 8 to explain that.

TRY ON YOUR OWN

In the previous lesson, we plotted the weights of Bob's and Alice's mice. The difference in weight was evident, but this is only a visual inspection of the data. We need to produce some actual numbers that will summarize our observations. Today's exercise is simple and straightforward (as this lesson) but not less important than any other exercise in the book. For the weights you already plotted, find out:

- The means of Alice's and Bob's mice
- The variances of Alice's and Bob's mice

What are your thoughts about these numbers?

Next

You should be able to summarize your sample by reporting the mean value of your observations, and the variance as the average sum of squares. Are these good estimates of the actual population values? Samples and populations are not always directly comparable, and we need to find ways of inferring population values from our sample. That's precisely the topic of the next lesson.

3

Estimating Populations

In the previous lesson, we discussed how to summarize measurements from samples. But what we really want to know is what the population looks like. In this lesson, we will cover how to estimate the characteristics of a population from samples.

How Do Populations Look Like: The Normal Distributions

If you recall, a population is the whole group of things we are interested in. As we can't measure all population values, we measure a subset of things, called a sample, and then we infer the population values from the sample values. But before we can do so, we need to know what populations look like.

Let's suppose that you are measuring the lengths of the ants you have in your garden. You want to have a fair idea of what are the mean length and associated variance of your garden ants, and to do so, you pick ten ants and measure them. If you plot a histogram, as we described in Lesson 1, you may have something like that:

DOI: 10.1201/9781003398820-3

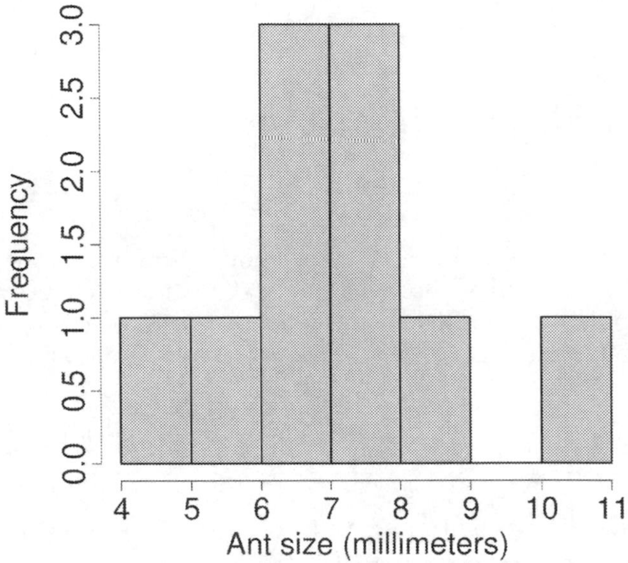

The data looks OK, but you feel you can do much better. Now you measure instead 100 ants, and the histogram of lengths will look like this:

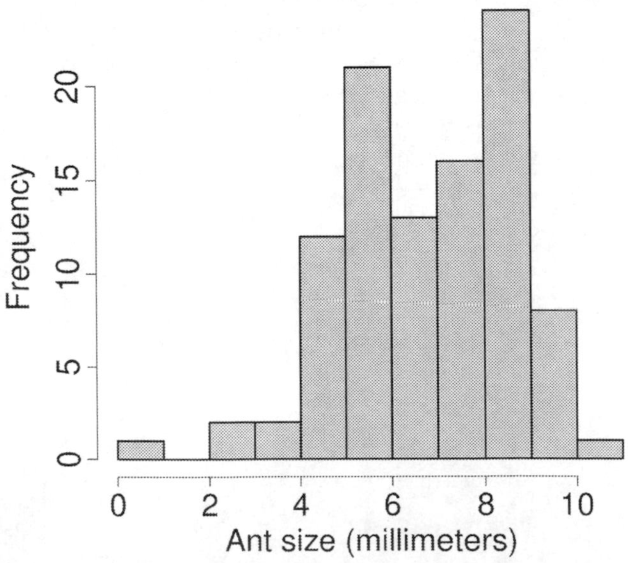

You can try with 1,000 instead, and that way, because you get much more data, you can reduce the size of your bins:

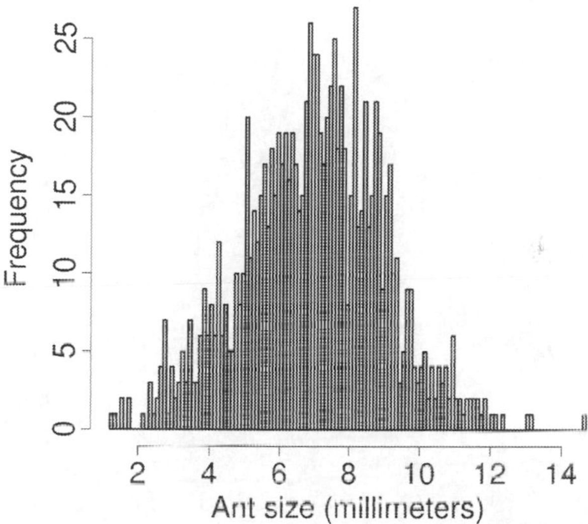

Now, you devise a machine that automatically sorts and measures ants from the garden, and you manage to measure one million lengths! What would the histogram look like? Something like this:

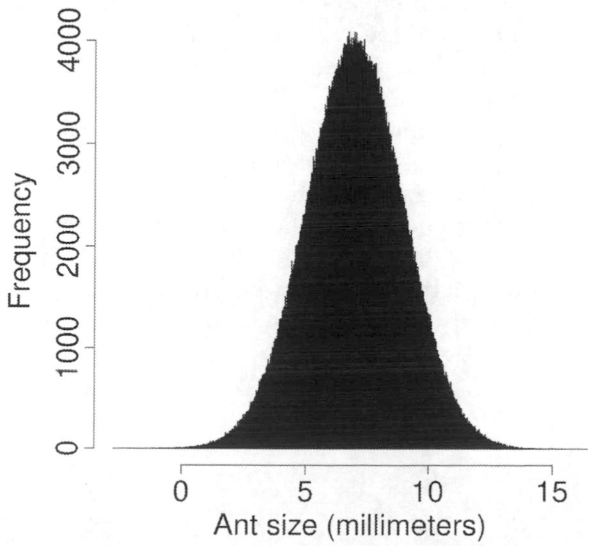

The shape of the histogram is quite distinctive, and its shape defines what we call a *normal distribution*:

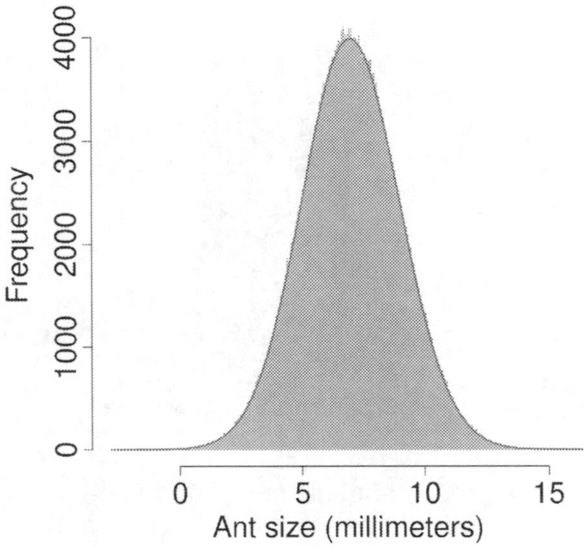

In nature, *most observations follow a normal distribution*. To define a normal distribution that corresponds to your population, you don't need a million numbers nor even ten, you just need two numbers: the mean and the variance (I bet you saw that coming). The mean defines where the centre of the distribution is, in our example, right at 7 millimetres. The variance tells us about the width of the distribution. The specific distribution with mean 0 and a variance of 1 is called a *standard normal distribution*. Our ants fit perfectly to a normal distribution with mean 7 and variance 4. In the graph below you can see how a normal distribution with the same mean but different variance has a different shape, and if the mean is different but the variance is the same, the distribution keeps the same shape but is centred in the mean value:

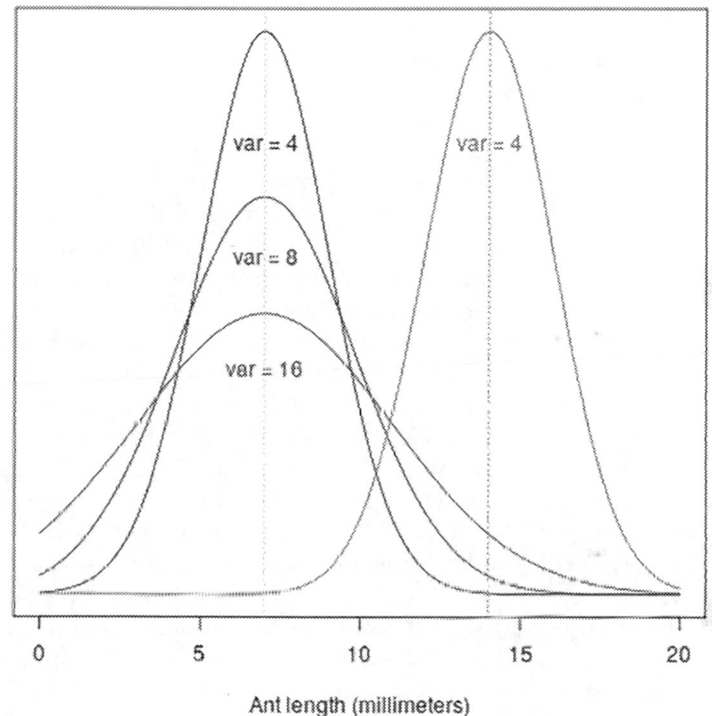

So, if we assume that our population is distributed like a normal distribution, you just need a small sample to estimate the mean and the variance, and you will have the whole population characterized!

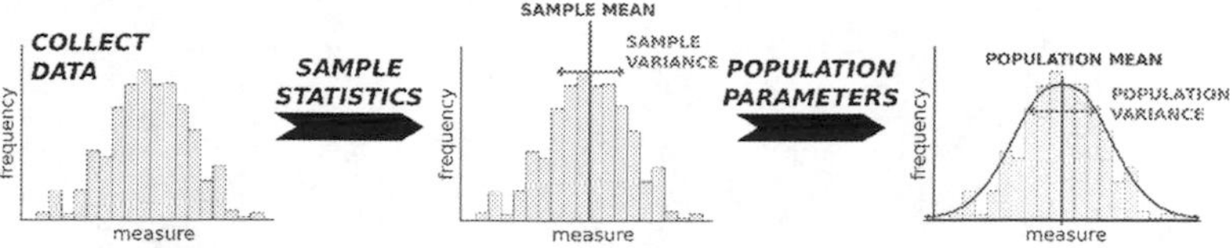

FIGURE 3.1
Populations as normal distributions. In statistics we often compute statistics of centrality and dispersion from the sample, and use them to characterize the underlying population structure by fitting a normal distribution.

Parameters and Estimates: Samples and Populations Redux

The values that define a population are called *parameters* that in the case of the normal distribution are just the mean and the variance. You never know for sure the values of the parameters. On the other hand, the values you compute from the sample are called *estimates* of the parameters. In this section, we will learn how to *estimate* the population *parameters*.

The easiest estimate is the mean (or average). It has been known almost from the beginning of statists that, if a population is normally distributed, the best estimate of the population mean is a sample mean. For instance, if in the example above the average of your lengths from a sample of ants is 7 millimetres, the best estimate of the population mean is also 7 millimetres. Easy peasy.

Notice though that when the sample size is very small, the estimation may not be good. For instance, if you sample two ants with sizes 5 and 7 millimetres, the average is 6 millimetres. You obviously need more than two data points to compute good estimators of the mean. In the graph below we show how the estimated mean approaches the true value (horizontal line) of the population parameters as we increase the sample size. This is called *the law of large numbers*:

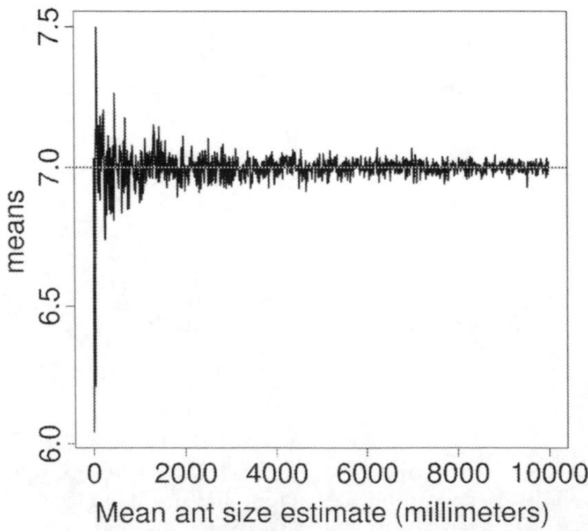

How about the variance? The answer is less obvious than you may expect. However, it will help us to introduce an important concept in statistics that we will discuss several times in this book: *degrees of freedom*. Let's start with a simple example. From the previous section, you know that the population of ants has a mean length of 7 millimetres and a variance of four. If you sample just three specimens, you may have the following measures (in millimetres): 5, 7 and 9. The mean of the sample (estimate) is 7 millimetres, which is exactly the population mean (parameter). However, if you compute the variance, that will be (on average) smaller than the population parameter. Specifically, the sum of squares in this case is eight (you can try on your own) so the variance (average sum of squares) is less than three, a poor estimate of the population variance.

The reason why the mean sum of squares of the sample is not a good estimator of the population variance is that we are getting the wrong type of mean: we are dividing the sum of squares (eight) by the sample size (size). But actually the number of measurements that give us some information is actually two. Why? In this case, if you are given the average (seven) and two of the measurements (five and seven), you can find out the third value, as only if it is nine, the three sample values will have an average of seven. In technical terms, we say that we have two degrees of freedom (as the third value is not free, and it's determined by the mean and the other two values). In this example, the mean sum of squares is computed using only two degrees of freedom, and the estimated variance is exactly four, the population parameter.

In general, to estimate the population variance from a sample, we always lose one degree of freedom. The example above was tailor-made to show the point. But if you compute the variance from three samples many times, you will observe that as you sample more and more the estimate using two degrees of freedom (line on top) converges to the true population value (again, using the law of large numbers) while if you divide by three (line in bottom), you underestimate the population variance:

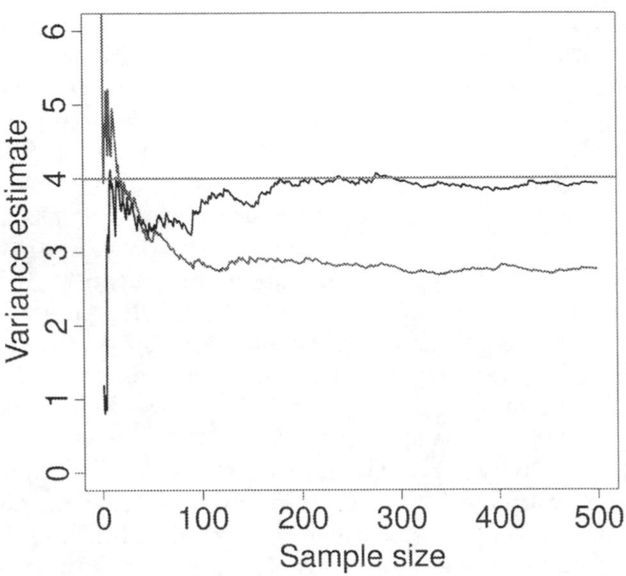

When an estimator converges to a value as the sample size increases, as in these two cases, we say it is a *consistent* estimator. However, when an estimator is systematically wrong, we say it is a *biased estimator*. The mean sum of squares of the sample is a biased estimator of the population variance. Instead, when an estimator is expected to give the correct value, we say it is an *unbiased estimator*. The mean sum of squares with one degree of freedom less than the sample size is an unbiased estimator of the population variance. (This last sentence seems complicated but it is really not, yet it is important. Please read it a couple of times making sure you understand the concept before going to the next paragraph.)

Degrees of freedom is one of the most confusing concepts in statistics so, if you are still a bit unsure, you are not alone. However, it is a very important concept and we will be computing degrees of freedom several times in this book so, don't give up. We'll re-explain degrees of freedom from a different angle in later lessons.

Expected Distribution of Squares: Chi-Squared Distributions

We talked earlier about the distribution of values in populations, mostly following a normal distribution. But as you may anticipate, we will do statistical analysis working with squares. So, how are the expected populations of squares distributed? The answer is the *chi-squared distribution*. The chi-squared distribution is to statistics what the normal distribution is to nature: it is the central distribution in statistical hypothesis testing, as we shall see.

Imagine a population of people attending a football match, and the arrival times are measured as the difference between the actual time and the scheduled start time of the event. That is, a negative time means that the person arrived early and a positive that arrived late. We assume that the times are distributed normally, with a mean time of 0 hours (i.e. right on time) and a variance of one hour (recall that this is a standard normal distribution). Now, pick a sample of just three people: one was five hours late, another was on time and a third was three hours earlier. We can compute the sums of squares graphically, as we did before:

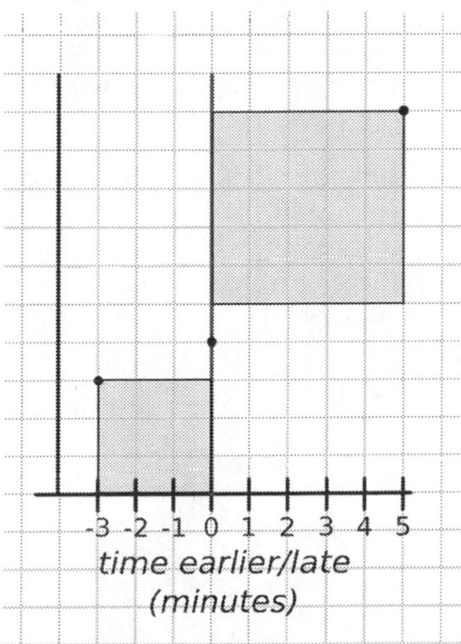

The sum of squares in this case is 34. Now, repeat this a million times (not you, we will let the computer do it) and build a histogram. How are the sums of squares distributed in this case? Here's the histogram showing the distribution:

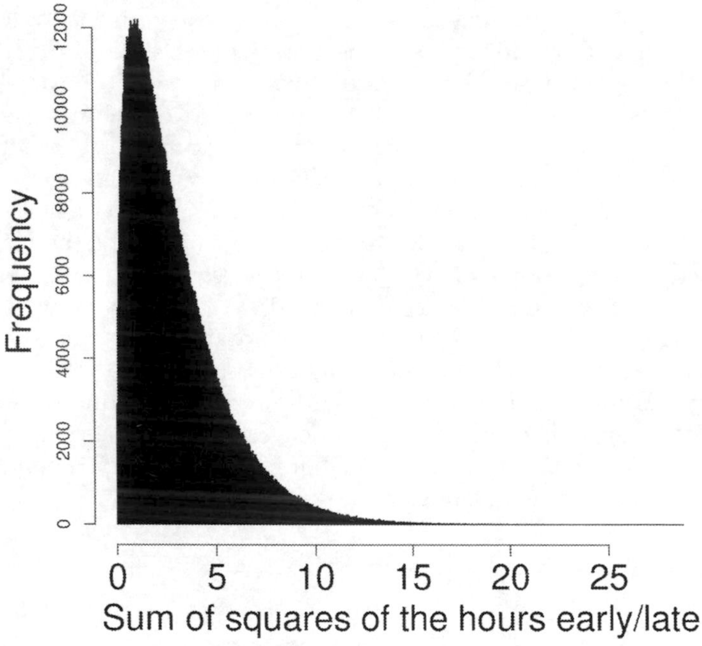

This is a chi-squared distribution with three degrees of freedom, that is, the number of values we used to compute the sum of squares. Indeed, we can define a chi-squared distribution to any number of degrees of freedom, even for one. Here's a graph showing the family of chi-squared distributions for different degrees of freedom:

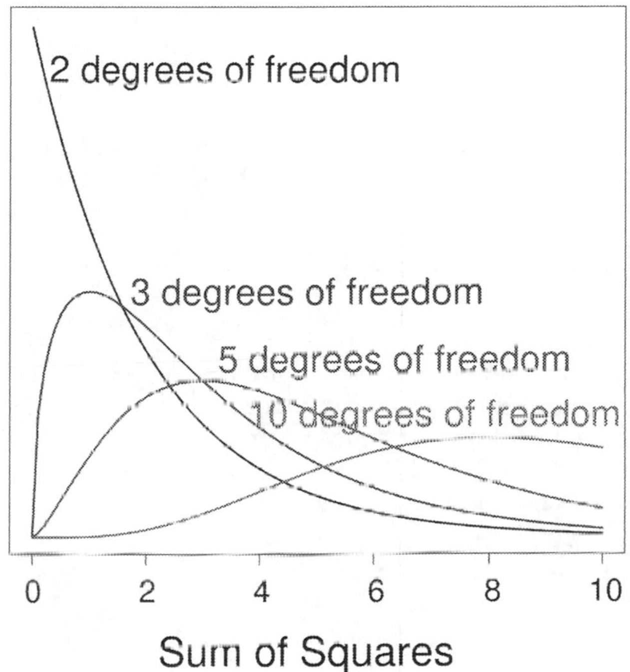

We'll be back to this distribution soon, but we can leave it for now. And so we conclude a long lesson full of important concepts. Time to do some practice.

TRY ON YOUR OWN

Alice doesn't admit that her mice are fat due to any diet. Bob is determined to show that diet has an impact, and he will estimate the effect on populations based on their small subset sample:

- Estimate the weight population mean and variance of mice in a fat-rich diet (Alice).

- Estimate the weight population mean and variance of mice in a fat-restricted diet (Bob).

Do you think that there is a noticeable effect of the diet on the mice weights?

Next

You now know important tools and concepts that you will be using to do statistical analysis. But before that, we need to establish the methodological framework that we have to use. In the next lesson, we will describe how samples are compared to obtain knowledge, and how statistical testing works.

4

The Design of Experiments

In Lesson 1, I introduced the concept of empiricism, or why science needs observations and experiments. How these observations and experiments are connected to knowledge is another matter of debate. Indeed, until relatively recently, we didn't know how science actually works (although again, this is still under debate between philosophers of science and scientists themselves). The first thing we need to describe here is how science works.

Hypothesis-Driven Science: The Deductive Approach

A naturalist observes that mammals (these furry animals around us) do not lay eggs. As a good empiricist, he takes more and more samples to be sure about his discovery. After sampling a large number of mammals, he can rightfully claim that 'mammals do not lay eggs'. But can he? There's a problem with this way of doing scientific inference, and is that we can never be sure of such a statement unless we sample all existing mammals in the world. Indeed, the naturalist will eventually find a platypus: a mammal that does lay eggs!

This sort of problem was already known to the empiricists, and it was called the *problem of induction*. Science must work in a different way. Knowledge should come from another approach. A solution to this problem is called the *deductive approach*. In this approach, we build upon previous knowledge to construct a *hypothesis*, and then, we do experiments or collect observations to *support* or *falsify* our hypothesis. As we discussed in regard to the induction problem, no matter how many observations we collect, we will never be totally sure of our hypothesis. However, if we show evidence about the hypothesis being wrong we will advance in knowledge as we will need to propose a new, and better, hypothesis.

In our example, the naturalist proposes the hypothesis 'mammals do not lay eggs' based on his knowledge. Then, he goes into the wild and starts to sample more mammals that will potentially support his hypothesis. However, he discovers the platypus! His hypothesis has been falsified. Far from being angry, he takes this opportunity to propose a novel and better hypothesis. He noticed that the platypus is not quite like the other mammals: they have an extra bone in their shoulders called the coracoid, like those in birds. His new hypothesis is: 'Birdy mammals lay eggs, non-birdy mammals do not lay eggs'. He goes into the wild again and finds other mammals without the coracoid bones that don't lay eggs. But he also finds the echidna, a mammal with coracoids (i.e. a 'birdy' mammal) and indeed it lays eggs! The new hypothesis is supported.

In science we don't prove things right, but we may show that a certain hypothesis is likely to be wrong. If a hypothesis holds after many experiments, we don't say our hypothesis is right, instead we say that our hypothesis is *supported by the data*. Remember, even if our hypothesis is great and robust, there can always exist a platypus that will falsify the hypothesis. And that's precisely how science works, and what determines how we do statistical analysis.

Interestingly, the use of hypotheses means that you are constraining your capacity to obtain knowledge by your intellectual ability to generate hypotheses. This connects in a way empiricism with rationalism. Indeed, Immanuel Kant (a rationalism champion) already wrote about these limits of knowledge. That's why many great scientific discoveries came after the proposal of novel, imaginative, even wacky hypotheses. Indeed, the inductive approach described earlier may not be good for testing hypotheses, but it is often used to generate new ones. That's why the scientific method is more about questions (hypotheses) than about answers. To propose a good hypothesis, you may need knowledge, intuition, observations and insight. To test a hypothesis, you need statistics.

The Design of Experiments

A common misconception is that statistical analysis starts after the data has been collected. Actually, statistics starts way before you collect data or perform any experiments.

The first step is proposing a hypothesis. Obviously, this hypothesis comes from previous knowledge so there's an even earlier step, which is getting this knowledge. But this can be considered part of your hypothesis design. For instance, literature search, pilot studies (induction style) or just thinking about a problem (rationalism style) is part of proposing a hypothesis.

Then, you decide on which sort of statistical analysis you are going to do. Are you going to compare two samples? More than two? Are the samples paired? Etcetera. After you decide your type of analysis, you start collecting the data. Finally, you perform the statistical test you already knew you were going to apply.

In summary, these are the steps to follow:

1. Set a hypothesis.
2. Decide on the statistical test based on your hypothesis.
3. Collect data (samples).
4. Estimate population parameters (statistics).
5. Test your hypothesis (using a statistical test).

There's still a further complication that we have to take into account. If you recall, we said that hypothesis cannot be proven right. So, how do we evaluate hypotheses? Statisticians here use a very neat trick: the evaluation of a *null hypothesis*. A null hypothesis is the hypothesis that 'nothing happens'. For instance, if you want to evaluate the hypothesis 'hares are faster than tortoises' your null hypothesis will be 'hares are as fast as tortoises'. If you show that the null hypothesis is not supported you will have support for your *alternative hypothesis*. So, in statistics, we don't prove hypotheses right, we may *reject* the null hypothesis and therefore find *support* for our alternative hypothesis. Keep also in mind that there is only one null hypothesis while there are infinite alternative hypotheses (hares are 1 mph faster than tortoises, or 2 mph faster, or 1.345 mph faster, etcetera) so, working with a simple well-defined null hypothesis makes things not only easier, but doable.

Comparing Hypotheses

As we shall see, statistical testing is about comparing your hypothesis (alternative hypothesis) with the hypothesis you would expect to be correct by chance (null hypothesis). If you recall from Lesson 3, natural populations are usually distributed like a normal distribution, and this distribution is defined by only two parameters: the mean and the variance. Let's go back to our analysis of the speeds of hares and tortoises. You can compile the data from a sample of each species, and then estimate the mean and variance of a population in which there will be no speed difference. If you plot the normal distribution associated with these values and compare it with the actual distribution of your two individual samples, you may observe that the samples' distributions and the expected distribution do not match (Figure 4.1).

Clearly, the speeds of the two species are pretty different and do not correspond to the speed distribution of the expected normal distribution, which is actually your null hypothesis. But this is a clear-cut case and other, less clear cases, can be found (Figure 4.2).

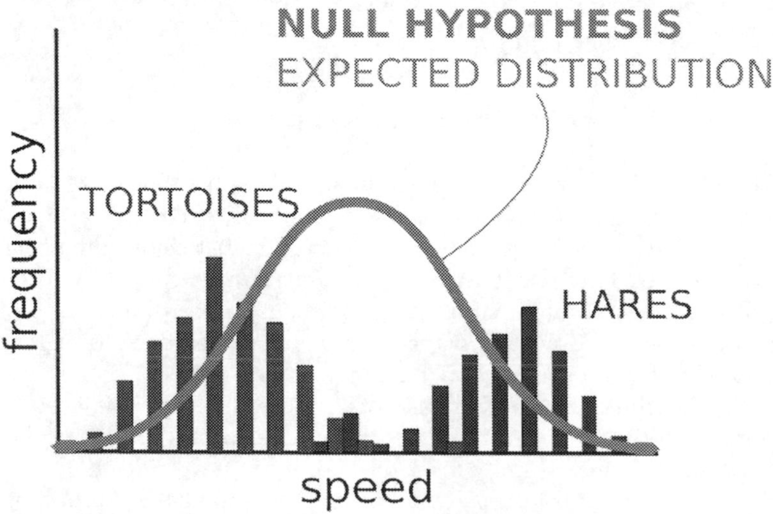

FIGURE 4.1
Fitting a normal distribution to two different samples. The distribution of the tortoise and hare speeds samples does not match the expected distribution under the null hypothesis.

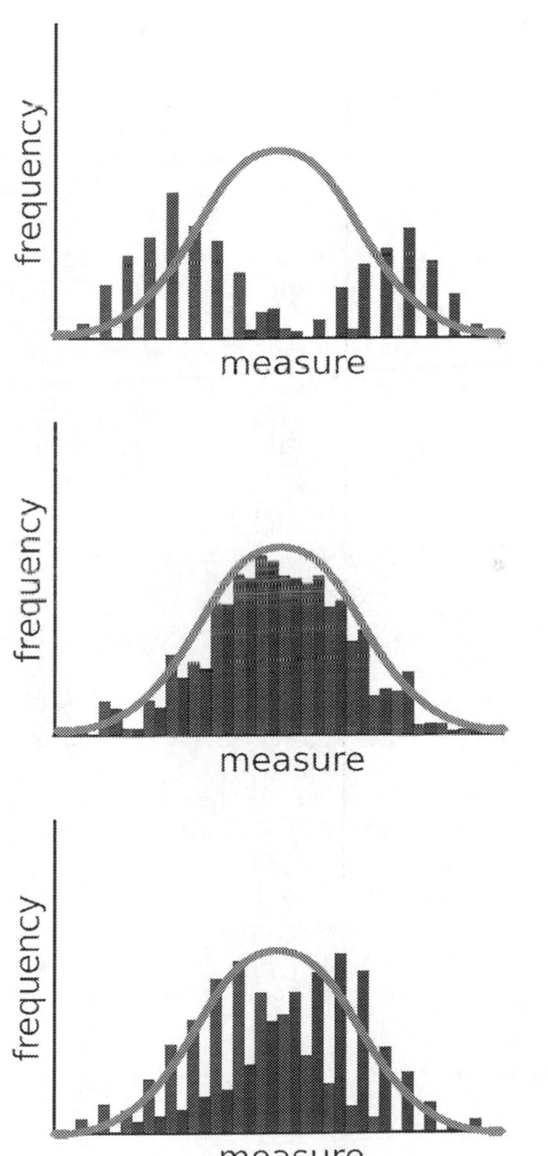

FIGURE 4.2
Various matches between the samples' distributions and the expected distribution under the null hypothesis. The match between samples distributions and the expected distribution can be poor (top), good (middle) or ambiguous (bottom).

The middle case in Figure 4.2 probably indicates that the speeds are not different to each other as they fit well with the null (expected) distribution. But how about the bottom case? Would you be confident to say that the speeds are different on average? We need a way to quantify this observed discrepancy between your expectations (the null hypothesis) and your expectations (the alternative hypothesis).

To compare hypotheses, we use in statistics something called a *P* value (although at the end of this lesson, I will briefly describe other alternative approaches to statistical analysis that do not rely on *P* values). A *P* value, or simply *P*, is the probability that, if you reject the null hypothesis, you are wrong.[1] This value ranges from 0 to 1 but can be expressed as a percentage. That is, a *P* value of 0.23 is equivalent to a 23% probability. For instance, if I tell you that for hares and tortoises you reject the null hypothesis (and hence support the alternative of different speeds) with a *P* value of 0.01, it means that with such a claim you still have 1% chance of being wrong.

In another experiment, you measure speeds of two species of hares and the null distribution of equal speeds is rejected with a *P* value of 0.25. That is, if you affirm that the two species of hares run at different speeds you have a 25% probability of being wrong. Would you run the risk? For traditional (and largely arbitrary) reasons, we often use a threshold of 5%. If you don't reject the null hypothesis with a *P* value smaller than 0.05, you won't express support for the alternative hypothesis. But for some applications, this is not enough: if you study a drug that does not kill the patient with a *P* value of 0.01, you can still kill potentially 1% of the population. Think of *P* values as a currency for statistical robustness, the smallest the better. After all, we are always wrong to a certain degree, but we want to control our chances of being wrong.

How to Compute *P* Values

To compute *P* values we make a number of assumptions, but mainly two:

- Your populations are *distributed normally*.
- Your samples are *independent* to each other.

By independent, we mean that the values from one sample are not influencing the values of the other sample. For instance, if you want to compare the speeds between tortoises and snails raiding tortoises, the speeds of the snails will depend on the speed of the tortoises below, hence, these speeds are not independent. A more realistic example: you want to test whether tall people accumulate more fat than short people. If you compare height and weight, you will be comparing two non-independent variables as the taller you are the heavier you are expected to be. However, if you instead compare height and relative fat content, you may be better off as they are likely to be independent of each other (kind of).

It's time to go back to the very first example in the book, that of the hours of sleep between young and old people. If you go back to the first few pages, you will see that the sum of squares under the null hypothesis of no difference between the hours of sleep was eight, and the sum of squares for differences (here or alternative hypothesis) was four. However, as the sleeping hours are normally distributed, we expect that by chance, sometimes, the sum of squares under the null hypothesis would be four or even less. But how often?

We start by characterizing the normal distribution that we would expect under the null hypothesis. This normal distribution will have a mean of 7 (average of the four data points) and a variance of 8/3 (the sum of squares divided by the degrees of freedom). Using this normal distribution, we can randomly get a sum of squares from a chi-squared distribution as expected under the null hypothesis. Let's get 100 random values (we are using a computer here):

```
16  4  5  7  5 13  5 10  2 32
 4  7  5  8 28 17  9 15 10  7
15  1 11  8  2  2  5 15  7  3
 9 26  1 10  9  3  4 12 10  7
11  0  1  2  6  4 24  5  4  9
 6 10 15 19  2  6  8  1  4  4
 0  6 13 10 32  8  1  8 20  6
 9 18  2  7  6 15  7  5 13  6
 1  9  4 14  5  2 30  3  7  9
17  1 16  1 16  9 11  5  3  7
```

We can now count how many sums of squares are equal or even smaller than the sum of squares of the alternative hypothesis, that is, four (in bold):

16	**4**	5	7	5	13	5	10	**2**	32
4	7	5	8	28	17	9	15	10	7
15	**1**	11	8	**2**	**2**	5	15	7	**3**
9	26	**1**	10	9	**3**	**4**	12	10	7
11	**0**	**1**	**2**	6	**4**	24	5	**4**	9
6	10	15	19	**2**	6	8	**1**	**4**	**4**
0	6	13	10	32	8	**1**	8	20	6
9	18	**2**	7	6	15	7	5	13	6
1	9	**4**	14	5	**2**	30	**3**	7	9
17	**1**	16	**1**	16	9	11	5	**3**	7

We count that 29 out of 100 random values are 4 or less. That is, if we reject the null hypothesis, we can be wrong 29% of the time. In other words, our P value in this specific test is 0.29. In the following lesson, we will be using pre-computed tables to find out P values.

On Being Wrong

Statistics, as we discussed extensively, is about dealing with uncertainty. In other words, you are never 100% sure of a hypothesis, but you can quantify how likely you are to be wrong. In this lesson, we talked about P values as the probability of being wrong if you reject the null hypothesis. We also call a P value the probability of a *type I error*, or a *false positive*. Conversely, if we don't reject the null hypothesis but the null hypothesis is wrong, we are having a *type II error*, or a *false negative*. In this book, we will not use this nomenclature nor will we compute the number of false positives or false negatives (neither true positives nor true negatives), but you may find this paragraph useful to revise if you find these terms in the literature. Also, these are extensively used in a field within statistics that goes beyond the scope of this book: *machine learning*. But let's not get there (not in this book, at least).

'Frequentism', 'Likelihoodism' and 'Bayesianism'

I have to be honest with you at this stage. The statistical procedure presented in this book is a specific approach to statistical testing: the frequentist approach. It is the most popular, most common, most used and (in my opinion) the most logical way of doing statistical analyses. But there are other approaches. Some statisticians prefer to evaluate all possible hypotheses (if possible) and select the most likely: this is the likelihood approach. In reality, the likelihood approach has been already adopted by frequentists: we use a maximum likelihood criterion to find the alternative hypothesis (we will see this in the next lesson), and then evaluate the maximum likelihood hypothesis against the null hypothesis.

Another school of thought is that of Bayesian statistics. In simple words, in a Bayesian framework, a statistician finds the most appropriate hypothesis, and then provides a probability for the hypothesis to be correct. Notice the slight yet important difference in approach: Frequentists find a P value, the probability of your null hypothesis being supported by the data, while Bayesians find the probability of the alternative hypothesis being right. A Bayesian approach sounds more appropriate and straightforward, but there are technical and even philosophical difficulties with such an approach. We won't be covering this here, but keep in mind that other statisticians may disagree with the way this book teaches statistics, and this will mainly be the reason why.

Do Scientists Only Test Hypotheses?

The fact that science progresses via hypothesis testing doesn't mean that science is all about hypothesis testing. What I mean is that many scientists spend vast amounts of time and resources in the exploratory step, producing big amounts of data and producing knowledge that can be used to generate novel hypotheses. This is science too! However, there is an increasing tendency to believe that by generating data alone without testing hypotheses, we can make a substantial progress in science. Some scholars call this data-driven science, but it is actually a euphemism for 'induction'. I call it personally the Data-driven Science Delusion. It is actually a fascinating topic that has engaged scientists and philosophers, and that is way beyond the scope of this book. But if you decide to learn more about statistics, be ready for this sort of thinking. But my honest advice is, to be safe, make sure you work with hypotheses. After all, we know that induction (or data-driven science) doesn't work, and we have known this for way over two centuries.

TRY ON YOUR OWN

Before you started measuring the weights of Alice and Bob, you already had a clear design in mind based on your previous knowledge. Can you clearly state the hypothesis you want to consider (null and alternative) in your statistical test?

Next

You don't really need to generate random values to do basic statistics. The chi-squared and related distributions have been fully characterized so you can contrast your sum of squares with already available tables and graphs. How *P* values are computed is the subject of the following few lectures.

Note

1. Technically, it is the probability of obtaining results as extreme as those you have if the null hypothesis is correct. However, the re-arranged definition given in the main text is, in my opinion, easier to grasp.

5

Comparing Two Variances

In previous lessons, we described what a statistical test is, but we haven't done any yet. We are getting there, but let me introduce you one more lesson before we go into statistical testing practice. Here we are covering the general methodological framework that we will be using in most statistical tests so, please, pay attention.

The first approach when one starts studying statistics is to compare the average of one sample with the average of another sample, and then compute a *P* value. However, this approach has a limitation; it does not work if you want to compare more than two samples. Even more, what if you want to compare several samples, some of which are related and some others are not? We could start with the simplest methodology and then introduce novel concepts every time we introduce a new test. But our approach here is going to be different: we will start with a general framework that, once you understand, will be applicable to most other statistical tests, no matter how complicated they are. The trick is, instead of comparing average values, we will be comparing variances.

This lesson is going to be more conceptual and less practical than the previous ones, but I still promise that no formulas will be shown. From now on, and I hope you forgive me, we are going to deal almost exclusively with variances. Averages are informative, but variances are statistically powerful!

Sums of Squares Around Any Value

Let's recall the example we used to compute our first variance in Lesson 2. We started by computing the sum of squares between the given values and the estimated average (four):

DOI: 10.1201/9781003398820-5

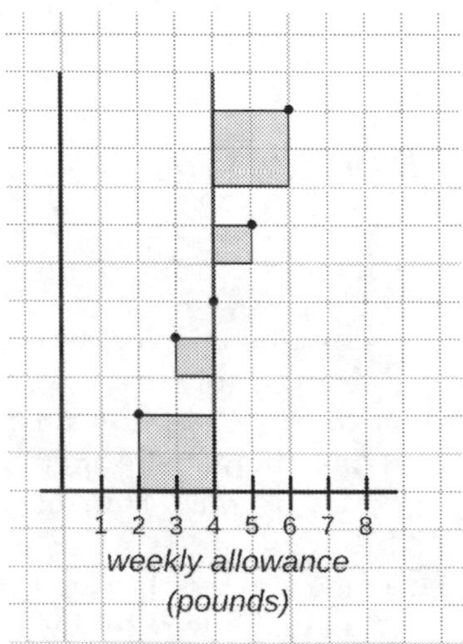

Here the sum of squares was ten. If you have a look at a mathematical statistics book, the variance is defined as the *second moment around the mean,* which in plain terms is like saying the *average of the square of the differences between values and the mean.* Does it mean that we can compute sums of squares around other values different to the mean? Of course!

Take some graph paper and compute the sum of squares around the value 5:

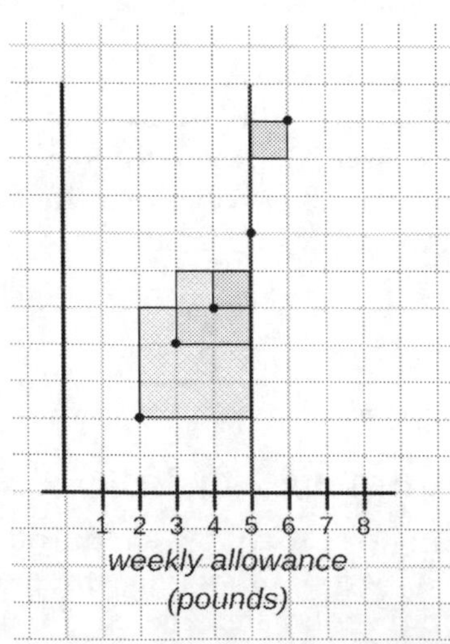

In this case, the sum of squares around the value 5 is equal to 15. How about around 6?

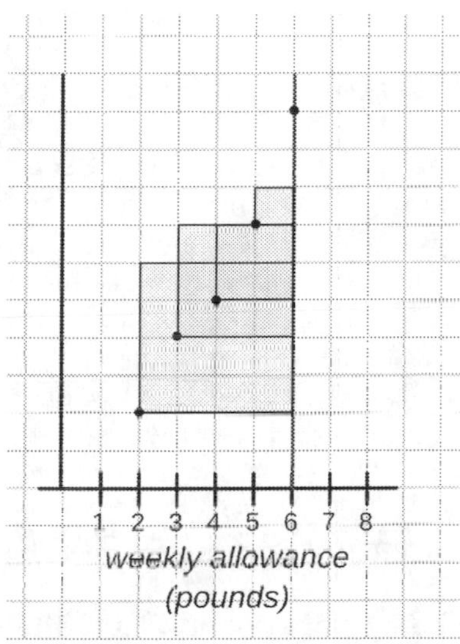

This sum of squares is equal to 30. You should notice that the farther away we go from the estimated average, the larger the sum of squares. Obviously, we don't know the real average of the population, as we are estimating it from the sample, so our estimate may be close or not that close to the real value. We know (or assume) that the population values are normally distributed, hence, the sum of squares will be distributed as a chi-squared distribution (as we described in previous lesson). The sum of squares around the estimated average (four in our example) represents the *error* around the estimated average. The sum of squares around a given average is called the *total* sum of squares. The difference between both sums of squares (total minus error) is called the sum of squares of the *model*. We will not go into details about these yet, as it will be explained in the context of the first statistical test in the next lesson.

In our example, let's suppose that you want to evaluate the hypothesis that the average value of the population is 6, for which we compute the *total* sum of squares as 30. However, the most likely hypothesis according to our data is that the average is four, which yields an *error* sum of squares of ten. The difference, that is 20 (30 – 10), is the additional sum of squares needed to explain the *model* (i.e. our hypothesis).

You may have noticed that the sum of squares for the estimated average (error) is the minimum value, and computing any other sum of squares will be bigger. Indeed, the *error* sum of squares is called the *least squares*. Remember when in previous lesson I said that something called *maximum likelihood* is used to find the appropriate alternative hypothesis in a test? Well, if the samples are normally distributed and independent, the maximum likelihood estimate of the average is precisely the one producing the least squares. In other words, *the model sum of squares is associated with the null hypothesis*, and *the error sum of squares is associated with the alternative hypothesis*, as we shall see in the next lesson.

If we can somehow compare the relationships between these sums of squares, or their associated variances, we will be able to provide a *P* value to test our hypotheses. To do so, we need to introduce (not yet) another statistical distribution.

The *F* Distribution: Comparing Two Variances

In previous lessons, we introduced the normal distribution, and also the chi-squared distribution which is in turn the distribution of sums of squares from a standard normal distribution (please revise if you don't recall this). From the sum of squares, you can compute the associated variance estimate by dividing it by the number of degrees of freedom. These variances are distributed as a chi-squared distribution but a bit flat, this is called a *scaled chi-squared distribution*, as shown in the graph below:

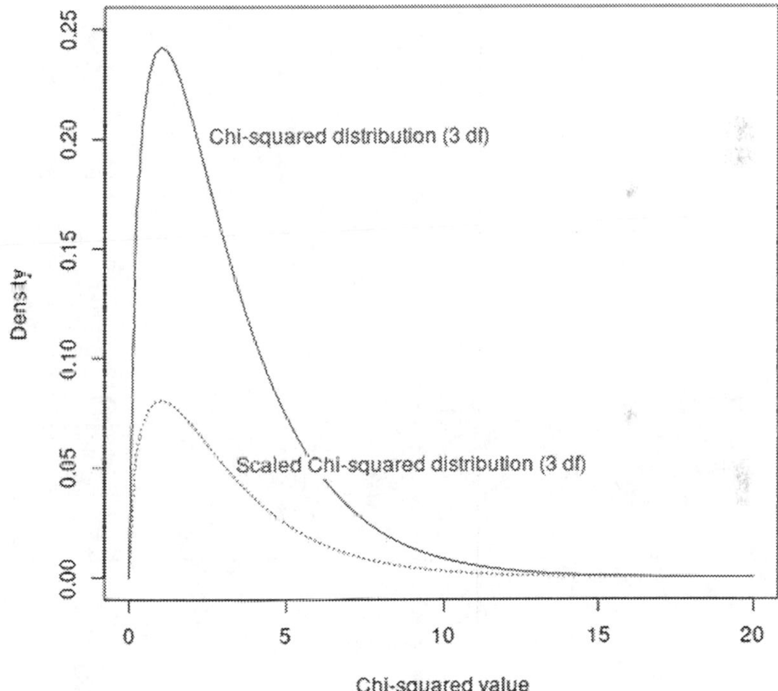

If you divide one variance estimate by another variance estimate, the resulting value will be distributed as something we call an *F* distribution (which is defined as the distribution of the ratio of two scaled chi-squared distributions). But don't be overwhelmed by the introduction of another statistical distribution, think of it as a combination of two chi-squared distributions.

You will recall that a chi-squared distribution is defined by the number of degrees of freedom so, how many degrees of freedom would the resulting *F* distribution have? It will have two types of degrees of freedom, corresponding to those of the underlying chi-squared distributions. For instance, an *F* distribution formed by the combination of two chi-squared distributions with 3 and

6 degrees of freedom, respectively, will have 3 and 6 degrees of freedom.[1] Here's an example in a graph:

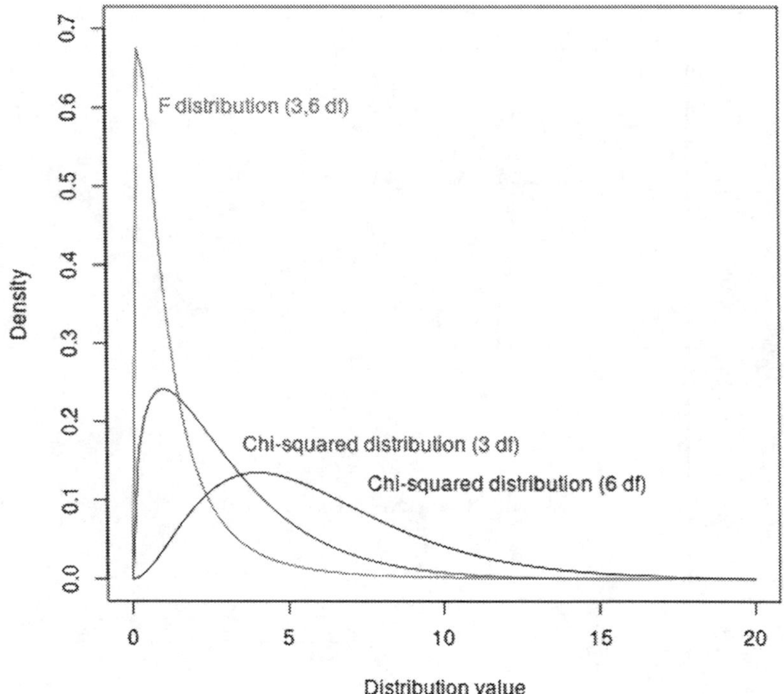

Now comes the useful bit. Let's suppose that we have two variance values (each distributed as a chi-squared). The first variance was computed using 4 degrees of freedom, and the second using 5 degrees of freedom (remember how variances are computed). If the two variances come from the same population, the ratio of the variances (first variance divided by second variance) will be distributed as an *F* distribution with degrees of freedom 4 and 5. What does this mean? That if the variance ratio is, for instance, 3, and the appropriate *F* distribution yields a *P* value of 0.13, there will be a probability of 13% of both variances coming from the same population (assuming normally distributed populations and all the other things we talked about). In other words, if our null hypothesis is that both variances come from the same population, we may reject it and support that both variances come from different populations, with a 13% chance of being wrong.

Now, please, stop for a moment and make sure you understand the previous paragraph. That's the basis of the general methodological framework to compare variances and perform statistical testing, as we should see now.

Analysis of Variance and ANOVA Tables

Classic statistics is usually explained around the concept of *Analysis of Variance* (or ANOVA in short). This is because most statistical tests are based on comparing variances (two or more). As you can guess now, the comparison of variances is done by computing variance ratios and finding out the probability of them coming from the same populations (of variances) via an *F* distribution. Which variances do we compare? Well, that depends on the type of test, and these variances are computed step-by-step, and their values together with the pre-computations are usually formatted in something called an *ANOVA table*.

In short, an ANOVA table contains a number of rows, one for each variance, and the columns are sequentially: the sum of squares (SS), the number of degrees of freedom (df) and the mean of the sum of squares, that is SS divided by df. Mean squares (MS) is another name for the variance, in other words, MS and variance are the same, but we often write MS for historical reasons. Last, we compute *F* values for each pair of variances we want to compare, in the simplest case, just one, as in the example below. Next to the *F* value, we can write the *P* value to be used in the interpretation of the test outcome.

In the example from the previous section, we wanted to test our hypothesis that the average value in our population was 6, and we computed a total sum of squares (SS) of 30, an error SS of 10 and, the difference was 20 as the SS of the model. We start by putting this value in the ANOVA table:

	SS	df	MS	F	*P*
MODEL	20				
ERROR	10				
TOTAL	30				

Next step, the degrees of freedom. We will revisit this for every test so don't worry if you don't understand exactly how they are computed right now. But the recipe is simple in this case. For the total sum of squares, we used five points, but we lost one degree of freedom (as we described in Lesson 3). A *linear model* (i.e. a straight line in our grid) requires only one degree of freedom to be defined. Last, as degrees of freedom are also additive (like the sums of squares), the error SS should have 3 degrees of freedom (4 – 1). Now, we can compute the variances (or MS) associated to the model and the error by dividing SS and df:

	SS	df	MS	F	*P*
MODEL	20	1	20		
ERROR	10	3	10/3 (3.3)		
TOTAL	30	4			

And finally the *F* value as a ratio of the variances and, if we get some help from a computer or a table, the associated *P* value:

	SS	df	MS	F	*P*
MODEL	20	1	20	6	0.09
ERROR	10	3	10/3 (3.3)		
TOTAL	30	4			

At the end of this book, there are a few *statistical tables* that can be used to find out *P* values or other computations of interest. Statistical Table 1 gives the *P* values associated with *F* distributions with one degree of freedom in the numerator (like the ones you will be computing in the next two lessons).

In conclusion, the model and the error variances seem different, but if we claim that they are different, we will have a 9% probability of being wrong, and that's quite a large probability. So, even if four is the most likely average of the population, we cannot easily discard that six is the true population average.

We will go through these steps in detail again in the next few lessons.

TRY ON YOUR OWN

Bob is determined to do an experiment that will settle, for once and for all, whether Alice's diet on mice is adding extra weight. While he writes the grant that will hopefully provide enough funds to do the experiments, he wants to do a preliminary analysis to convince the money givers (this is called a *pilot study*). Bob is about to build his first ANOVA table, and you are going to help him.

Build an ANOVA table to compare the variance associated with Bob's mice and their average weight, and the variance that results if we hypothesise that the population represented in Bob's mice have the same average weight as that computed from Alice's sample.

Next

Now you are fully equipped to not only do statistical tests but also understand what is going on in every step. Next stop: statistical tests to evaluate one-sample.

Note

1. For the 'maths' people: you may be worried by the fact that the underlying chi-squared distributions are not standard, as the population values do not necessarily follow a standard normal distribution. However, when you divide one chi-squared distribution by another one, both coming from the same normal distribution (even if it is not standard), the ratio is the same as if they were computed from standard normal distributions. Basically, the population variances cancel each other.

6

One Sample

This lesson will introduce the first statistical test, that is, a way to test whether your sample (or samples in later lessons) agrees or not with a specific hypothesis. In a previous lesson, we characterized a population of ants by taking a large number of samples and computing the average length. If you recall, we conclude that this population had a length normally distributed with a mean of 7 millimetres. Let's suppose that you discover a nearby colony and you wonder whether these ants have the same distribution of lengths. We will follow the general experimental design described in Lesson 4 (as we will do for all other tests presented in the book). For a one-sample test, this is the procedure:

1. *Set a hypothesis*: We hypothesize that ants in the new colony follow the same length distribution as the ants from the already known colony (this is our null hypothesis).
2. *Decide on the statistical test based on your hypothesis:* we will perform a *one-sample test for differences.*[1]
3. *Collect data (samples):* We decided to sample four ants from the new colony. The measured lengths in millimetres were 2, 4, 6 and 8.
4. *Estimate population parameters:* Following the instructions given in the previous lessons we estimate the population mean from the sample, assuming that the length in the original population follows a normal distribution. In our case, the estimated population mean is 5.
5. *Test your hypothesis* (using a statistical test): In this case, as we decided in step 2, we will find out our P value and support for our hypothesis by performing a one-sample test for differences. We will compare with our test the values of 5 (estimated) and 7 (our null hypothesis) via their associated variances.

DOI: 10.1201/9781003398820-6

One-Sample Test for Differences

Get a piece of graph paper and start by drawing two perpendicular lines, a vertical representing the sample values (ant length in our example) and a horizontal representing the samples (just one in this lesson). Then, draw a dot for each sample value:

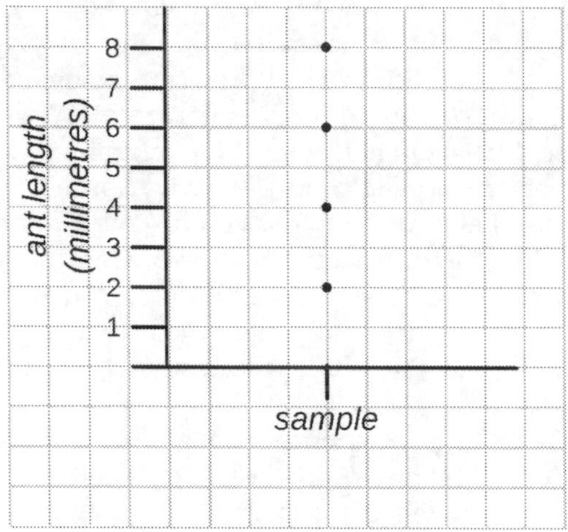

A horizontal line (red in the graph below) at the value we want to test (null hypothesis) is added to the graph:

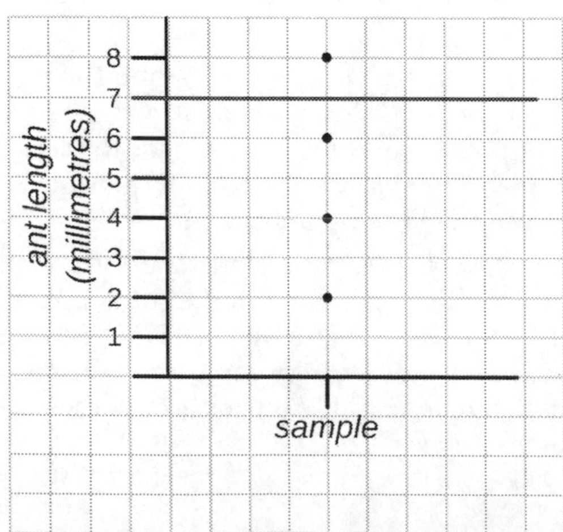

As we have done several times now, you can compute the sums of squares of the sample around the given average:

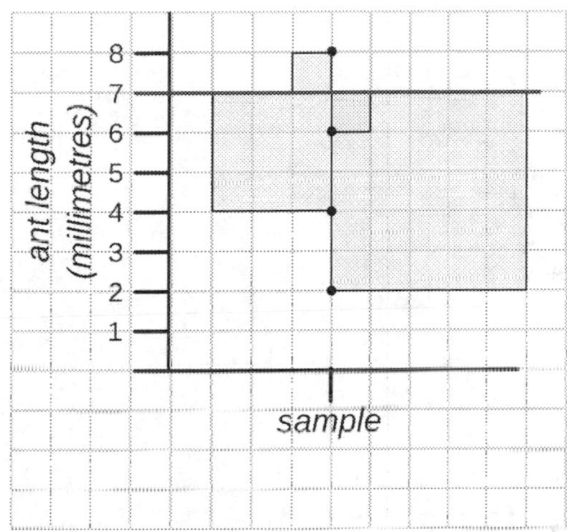

If you count the little squares enclosed in the bigger connecting squares, this adds up to 36, and it is our *total* sum of squares. You can start now populating your ANOVA table, remembering that the total number of degrees of freedom is the sample size minus one:

	SS	df	MS	F	*P*
MODEL					
ERROR					
TOTAL	36	3			

How about the error? First create a new plot, this time marking the estimated average (5 in our case):

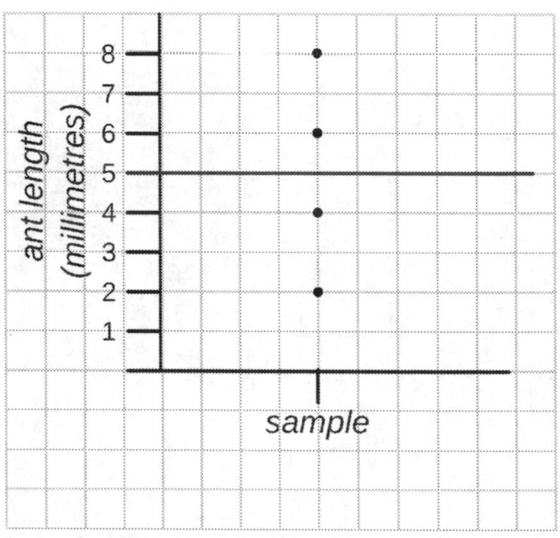

Then find the squares:

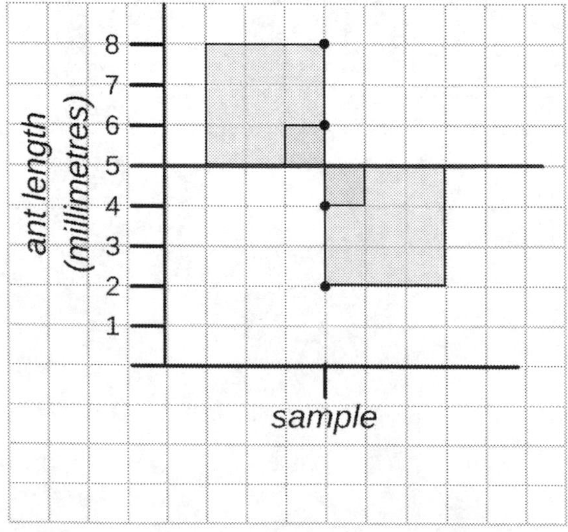

The so-called *error* sum of squares is 20 in this case.

How about the degrees of freedom of the model? The question is, how many numbers (items of information) do you need to define the model with respect to the hypothesis average? Just one, because you are moving the line horizontally by adding a single quantity (adding 2 to the sample mean, in this example). Hence, the number of degrees of freedom of the model is one, so the ANOVA table so far looks like this:

	SS	df	MS	F	*P*
MODEL		1			
ERROR	20				
TOTAL	36	3			

Using the property of additivity of degrees of freedom and sums of squares, you can fill the gaps in the table, and then compute the variances (recall, the mean squares or MS) and the *F* value. You can also find the *P* value in Statistical Table 1 at the end of the book:

	SS	df	MS	F	*P*
MODEL	16	1	16	1.6	0.293–0.423
ERROR	20	2	10		
TOTAL	36	3			

From Statistical Table 1 we can't compute an exact *P* value as it is only for integer values of F, but we know it is between 0.293 and 0.423. These are very high, high enough to conclude that we cannot reject the null hypothesis. Despite the estimated value of 5 being very different to our hypothesis of 7, we cannot discard, statistically speaking, that the average length of the ant population from which we took the samples is actually 7 millimetres.

Model Sum of Squares

Can we compute the sum of squares of the model directly, without relying on the additivity property? Yes, we can. Here's how. First create (yet another) new plot by drawing horizontal lines for the estimated and the hypothetical averages, and fit a square for their difference:

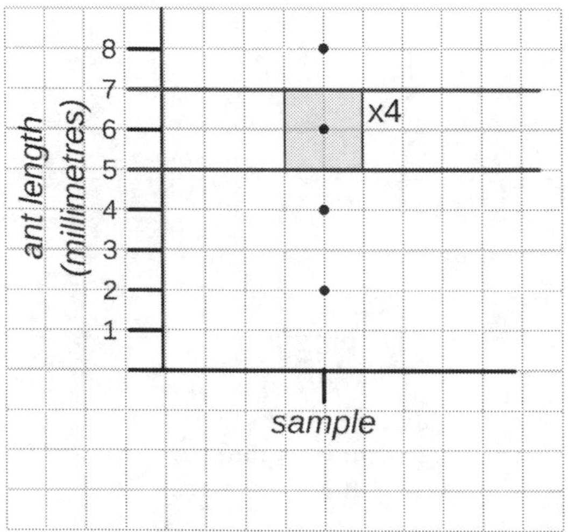

The sum of squares is the sample size times the number of little squares. Why? Because the sum of squares of the model is the sum of squares for each point in the sample of the difference between its estimated average and the hypothesis average, which is the same for all four points. Anyhow, the sum of squares of this model is 16 (four times four), as we already knew from the ANOVA table.

That should convince you already that sums of squares (and the degrees of freedom) have the additivity property, as we mentioned several times in the text. And this is precisely the reason why we use sums of squares and not sums of lines, or of cubes. We will explain this in detail in Lesson 8.

TRY ON YOUR OWN

Bob secured some small funding that allowed him to get five new mice to do experiments. Since this is a small sample, he put them all in a single group: mice under Alice's diet. So, he fed the four mice with a fat-rich diet for six weeks and measured their final weights, which were 24, 30, 28, 27 and 26 grams. His null hypothesis is that their average weight is not different to the average weight of his house mice: 18 grams.

Perform a one-sample test to evaluate whether the average sample weight is significantly different from the expected weight under the null hypothesis.

Next

The one-sample test is useful to get started in statistical testing, but it is rarely used. By far, the most used tests are those that compare two samples, as we will see in the next lesson.

Note

1. In classic statistics, this is known as a *one-sample t-test*.

7

Two Samples

In the previous lesson, we assumed that we knew the population average value, or at least we had some good evidence for it, and then we compared this value with the only sample we had. However, we usually don't know the population average. Also, if you want to examine the effects of a condition or treatment (diet, drug . . .) the more appropriate comparison is between a *treatment sample*, and an untreated sample (*control sample*) of similar sizes. For that reason, the most used (and often abused) statistical test is the one presented in this lesson.

The example for this lesson is a classical clinical trial. We are to test a new drug that increases the levels of insulin in the blood. To do so, eight people are recruited, and four of them will get the drug (*treatment group*) while the other four will get a placebo, that is, something that looks like the drug (pill shape, colour etc.), but it does not contain the actual drug (*control group*). Obviously, the patients don't know whether they got the drug or the placebo, and the assistants collecting the data don't know either (this is called a *double-blind experiment*). For a two-sample test, we proceed as follows:

1. *Set a hypothesis*: We hypothesize that the average insulin levels of both groups after taking the drug are the same. In other words, both samples (control and treatment) come from the same population: people in which the drug does not have a detectable effect (this is our null hypothesis).
2. *Decide on the statistical test based on your hypothesis:* We will perform a *two-sample test for differences*.[1]
3. *Collect data (samples):* We are measuring the insulin levels in microunits per millilitre (mcU/ml) in all eight participants after two weeks of taking one gram of the drug every day. For the control group, we got the values 1, 3, 5 and 7. For the treatment group, we got 7, 9, 11 and 13.

DOI: 10.1201/9781003398820-7

4. *Estimate population parameters:* We need to estimate three averages here. First, what would be the average of the population if all samples came from the same population? This will be the average value of all eight measurements, that is, seven. Next, what would be the average of each population if the two samples came from different populations? We can estimate this by computing the average of each group, that is, four for control, and ten for treatment.
5. *Test your hypothesis* (using a statistical test): Using the test described next, we will compare the expected average level of insulin assuming that both samples come from the same population (null hypothesis) with the average levels in each sample. And as we did in the previous lesson, we will be doing this by comparing their associated variances.

Two-Sample Test for Differences

We start again with a piece of graph paper, but this time we are going to plot the dots of two samples, the control and the treatment, as follows:

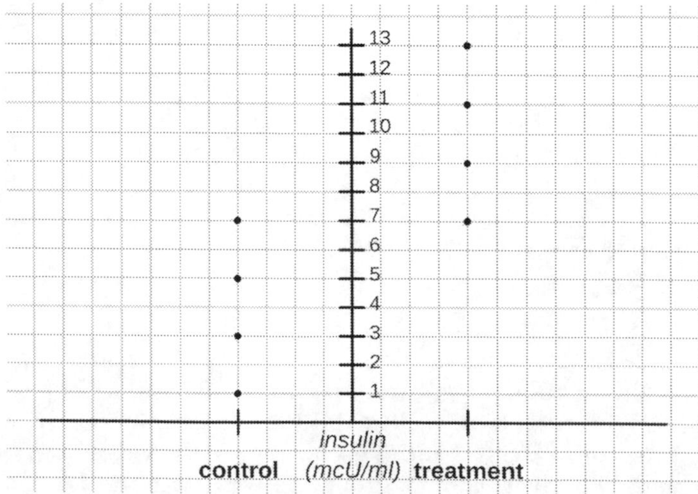

As in the one-sample test, we need to draw a line indicating the null hypothesis. On the one-sample test, we are given the value. In a two-sample test, we have to compute it. If, under the null hypothesis, the two samples come from the same population, they should have the same average. And what is the best estimate of the population average? The average of the combined sample, in our case, seven:

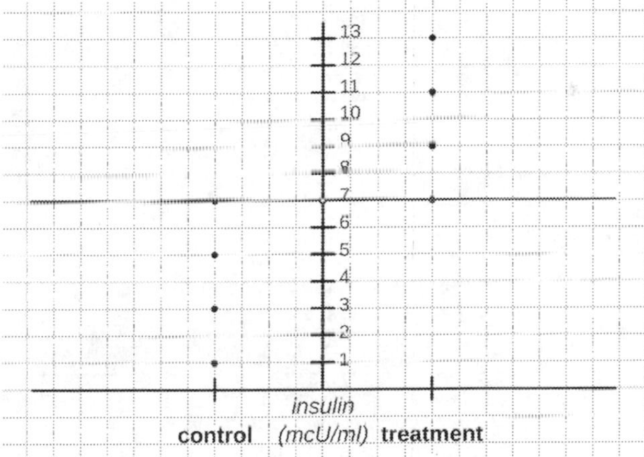

We can now start to connect sample values with the total average (our null hypothesis) and fit the square to compute the sum of squares:

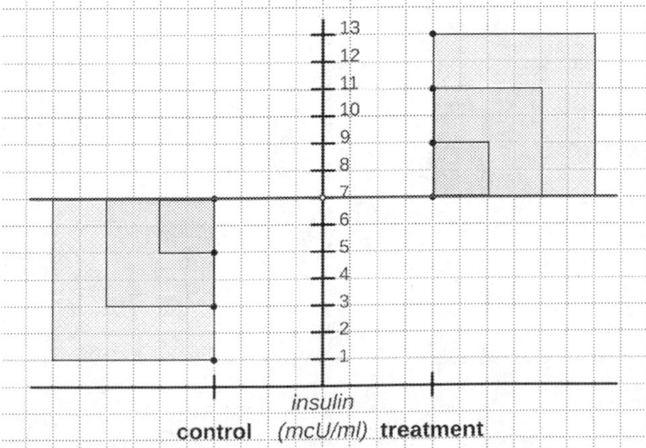

So, the total sum of squares is 112. Now, we must find a model representing the alternative hypothesis (remember, this is the least-squares model). In the one sample test, the line was parallel to the null hypothesis line. In a two-sample test, the line is a rotation of the null hypothesis line. We find it out by calculating the average values of each sample, 4 and 10 in this example:

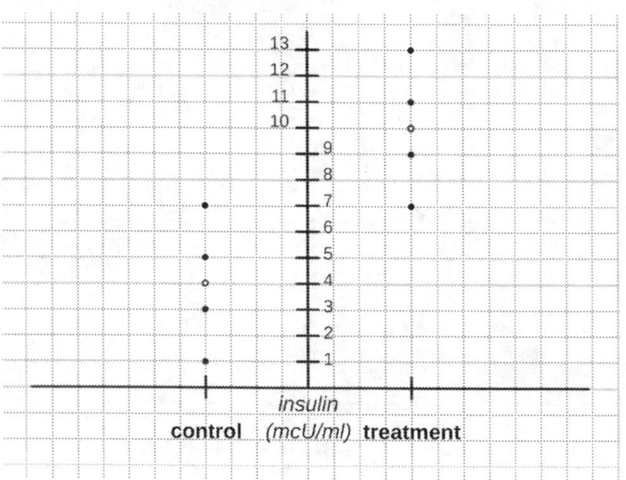

And then we connect the points with a line:

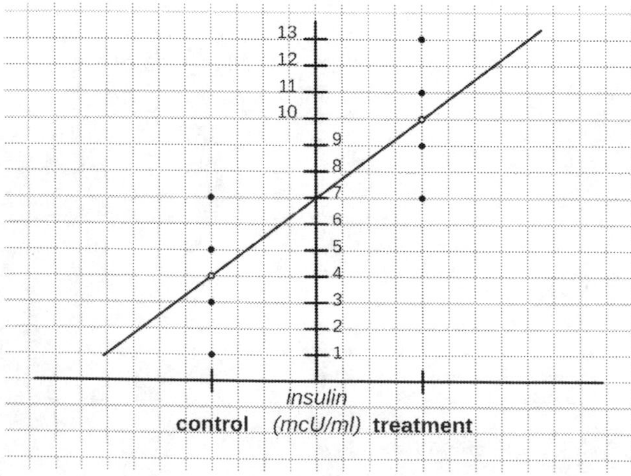

You should notice that the x-axis is intersected also at the combined sample average (seven). This is not by chance, and it is a property that we will exploit in more sophisticated analysis like regression, but we still have to wait a bit for that. Next, we find the squares and compute the error sum of squares:

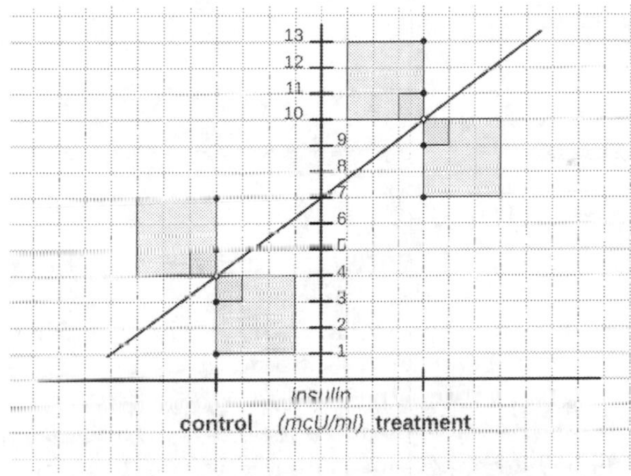

In this case, this is 40. We can start already to populate our ANOVA table:

	SS	df	MS	F	*P*
MODEL	72				
ERROR	40				
TOTAL	112				

Remember that the degrees of freedom are also additive. We have a total of eight values, so we have seven degrees of freedom in total. To define the model with respect to the total (sample), average we just need one number, the rotation angle of the straight line, hence we have one degree of freedom for the model, and therefore six for the error. We can finalize our table as follows:

	SS	df	MS	F	*P*
MODEL	72	1	72	10.8	0.016–0.020
ERROR	40	6	10		
TOTAL	112	7			

The *P* value according to Statistical Table 1 is between 0.016 and 0.020. With a computer, you may find out the exact *P* value, which is 0.017.

Using the definitions in this book, you would say that by rejecting the null hypothesis, you have 1.7% chances of being wrong. In plain terms, we say that the effect of the drug on the levels of insulin in the blood is *significant*.

One or Two Tails?

In this lesson's example, we rejected the null hypothesis with a *P* value of 0.017. That means that we supported the alternative hypothesis: the drug has an effect on insulin blood levels. But what we actually observe is that the drug has a specific effect on *increasing* the levels of insulin, as we expected. When we have grounds of evidence to believe that the average value from one sample will be higher (or lower) than that from the other sample, we perform a variation of our test called a *one-tailed two-sample test*. When we don't have prior information about the directionality of the difference, we do instead the *two-tailed two-sample test*, as described in the previous section.

A one-tailed test starts with a slightly different null hypothesis: the drug has no effect on insulin blood levels *or* it decreases the average level of insulin in blood. Therefore, the alternative hypothesis is: the drug increases the levels of insulin in the blood. This apparently mild variation makes our test more powerful, as the *P* value of the test will be half of that of the two-tailed two-sample test. That is, in our previous example, if we run a one-tailed two-sample test, we will obtain a *P* value of 0.0085.

Confidence Intervals

In standard statistical practice, it is often necessary to compute something called *Confidence Intervals* or CIs. Statistics is about quantifying the uncertainty about our experiments and observations.

So far, we have quantified this uncertainty by computing *P* values that tell us how confident we can be about our hypothesis. Another way of quantifying uncertainty is to estimate the most likely range where to find the true population mean values.

To do so, what one can do is to use the *F* distribution the other way around. First, you set a percentage of confidence, for instance 95%. This is the complement of the *P* value, that is, a confidence of 95% means that the *P* value threshold is set to 5% (and a confidence of 99% means a *P* value of 1% and so forth). Then the *F* distribution is used to estimate the maximum and minimum mean values that will fall within the set confidence (i.e. the values that do not fall beyond the *P* value threshold). But fear not, you don't need to do so. You can use a computer or, more conveniently, Statistical Table 2 provided at the end of the book.

In our example, you have a mean estimate of the control group and another of the treatment group, of 4 and 10, respectively. Each estimate was associated to a sum of squares of 20, and had a sample size of 5. From Statistical Table 2, you retrieve a value of 2.78. This is the value you have to add and subtract to each mean estimate to find out the 95% confidence interval. Hence, the intervals for our estimates are:

CI(95%) – control:	between 1.22 and 6.78
CI(95%) – treatment:	between 7.22 and 12.78

Notice that the confidence interval has two boundaries, a minimum and a maximum value. These quantities are related conceptually (and also mathematically) with the difference between one-tailed and two-tailed approaches that we discussed in the previous section.

Paired-Sample Tests

One common way to control for unwanted variation is to perform paired-sample tests. For instance, in the insulin example above, we selected eight random patients. In this case, it is difficult to control for sources of variation like the sex and genetic background of the patients, or even the diet or the environment. If, instead, we select four patients and measure the insulin level *before* and *after* the drug intake, we will have four pairs of values. In this case, it will not be appropriate to do the two-sample test described here, because we are violating one of the basic assumptions: independence of the samples. However, we can compute the difference in the insulin levels in each patient before and after drug intake, and do a one-sample test as in the previous lesson, evaluating the null hypothesis of no differences (i.e. a straight horizontal line right at zero). This is called a paired-sample test[2] and you can try on your own if you wish.

TRY ON YOUR OWN

A generous philanthropist, impressed by Bob's study, donated additional funds to include in his experiments a few more mice. Bob kept five additional mice under a low-fat diet, keeping otherwise the same conditions as those of the mice that were fed with a fat-rich diet. After six weeks, the weights were 19, 18, 21, 20 and 17 grams.

Perform a one-tailed two-sample test to evaluate whether the average weight is higher in mice with a fat-rich diet than those with a low-fat diet. Provide 95% confidence intervals for both mean estimates.

Next

Traditionally, you learn a simple way of testing two samples without using variances, but then you need to learn more concepts like variances, models and so forth, to approach more complex statistical tests. However, you did learn not only how to do the most popular statistical test but also the conceptual framework of more complex (and informative) tests, so this is clearly an advantage you have over other students that learn the so-called *t*-test classical approach first. But before we proceed with other tests, we will stop for a few minutes to finally explain why we use squares in statistics.

Notes

1. In classic statistics, this is known just as a *t-test*, in some places also called a *Student's t-test*.
2. In classic statistics, this is known as a *paired-sample t-test*.

8

Why Squares? (At Last!)

In the previous lesson, we filled our ANOVA table using the property of additivity of sums of squares and degrees of freedom, as already explained. We checked this in the one-sample test, let's do it now for the two-sample test. On a piece of graph paper, represent the two samples as in the previous lesson, and also the null and least squares lines, and add the connecting squares:

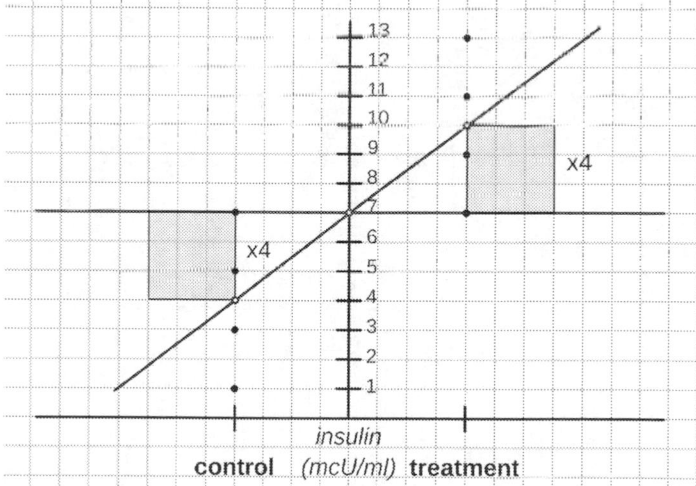

To count the sum of squares, remember that each square is represented four times in each sample. In this example, we see that the sum of squares of the model is 72, as we anticipated using the additivity property in the last lesson.

What if we use a different shaper other than squares? The first and most intuitive choice would be just the distance, right? OK, let's try out. We can repeat the whole two-sample test in the previous lesson, but in our ANOVA table, we will be recording the sums of distances (SD):

DOI: 10.1201/9781003398820-8

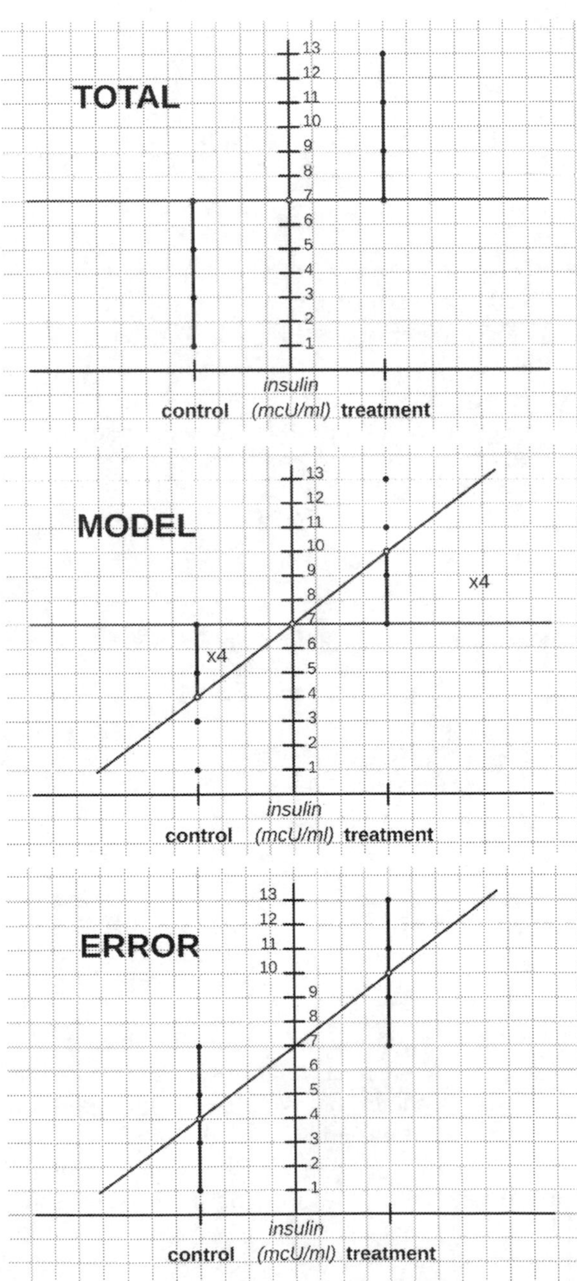
TOTAL
13
12
11
10
9
8
7
6
5
4
3
2
1
insulin
control (mcU/ml) treatment
MODEL
x4
x4
ERROR

The sums of distances will be:

	SD	df
MODEL	24	1
ERROR	16	6
TOTAL	24	7

Clearly, the total SD is not the sum of the model and total SDs. We can try other shapes, like cubes, or hypercubes.[1] Mathematically, we can define a shape factor that determines the area of the shape of an object (for some objects) in any dimension space, where a factor of 2 means a perfect square shape, a factor of 3 a perfect cube shape and a factor of 1 a straight line (distance), as in the previous example. If for the current example, we plot the difference between the total and the combines model and error for each shape factor, we observe this:

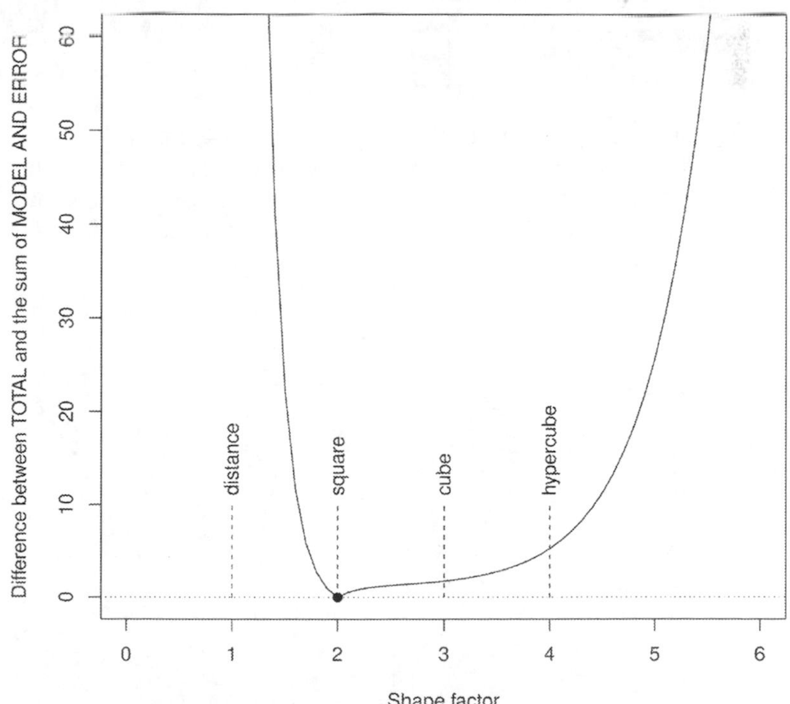

In other words, only for squares, the model and error values add up to the total value. That is, only squares have the additivity property in this context.

The additivity property of the sum of squares has been exploited in many ways in statistics, and in mathematics in general. Degrees of freedom, in this context, also have the additivity property, as we have seen in previous lessons.

TRY ON YOUR OWN

Bob is still unsure about the additivity property of sums of squares. Convince him by computing the sums of squares for the model and the error, and show that they add up to the total sum of squares.

Next

Often, computing a sum of squares is not easy, particularly when considering more than two samples. In these cases, we can use the additivity property to find sums of squares, as we will see soon. But first, we are going to generalize our tests for more than two samples.

Note

1. Cubes for more than three dimensions.

9

More Than Two Samples

Intuitively, if you have more than one sample, you could potentially do all the possible pairwise two-sample tests. However, there are two problems with this approach. First, the number of tests increases rapidly with the number of samples. For instance, for three samples, there are three possible pairwise tests, for five samples there are ten tests, for ten samples there are 45. Second, the larger the number of tests, the lower the overall power. Let me explain this with a little bit more detail.

If you run one statistical test and the resulting *P* value is 0.02, you have a 2% chance of being wrong if you reject the null hypothesis (as we have seen). But if you do ten tests at the same time yielding a *P* value of 0.02 each, the overall probability of being wrong is 2% combined for all ten tests together. Like in lottery games, if you buy ten tickets, you have more chances of winning a match. In statistical tests, the more you perform, the more chances you have of being wrong. For small *P* values, the overall probability of wrongly rejecting the null hypothesis is approximately the number of tests times the individual *P* values.[1] In this case, 10 times 0.02 is 0.2, that is, a 20% change of rejecting at least one null hypothesis! These sorts of computations are known as *multiple testing* corrections, and are very useful in fields where you need to perform many tests simultaneously (like in all the Big Data disciplines). I will explain a bit more about this in Lesson 14. But unless you have a special interest in doing so, the more efficient way of using your data for many samples is to combine all in a single test, producing only one *P* value.

By using the tests already described, we are now able to tell whether two samples come from the same population or not, and measure the uncertainty of our claims by computing a *P* value. But the general principles behind the sums of squares and the ANOVA tables that we used can be extended to more than two samples. Let's suppose you have four farms for which you suspect some may be using an illegal nutrient that makes chickens bigger

DOI: 10.1201/9781003398820-9

(but unhealthy). At this stage, you are mostly interested in whether chickens differ in size among the different farms. You weight 3 from each farm, totalling 12 chicken. For a multiple-sample test, we will proceed as follows:

1. *Set a hypothesis*: We hypothesize that the average weight of chickens in all four farms are the same. In statistical terms, the chickens from all farms come from the same population: the type of feed specific to each farm has no detectable effect (our null hypothesis).
2. *Decide on the statistical test based on your hypothesis:* we will perform a *multiple-sample test for differences.*[2]
3. *Collect data (samples):* We take the weight in kilograms of three chickens from each farm. The measurements were for Farm A: 6, 5 and 7; for Farm B: 3, 4 and 5; for Farm C: 4, 2 and 6; and for Farm D: 12, 10 and 8.
4. *Estimate population parameters:* We need to estimate five averages. First, the average weight of all samples came from the same population. This will be the average value of all 12 measurements, that is, 6. The other four averages correspond to the average for each farm, like if each farm had its own population of chickens. The average values for farms A to D are 6, 4, 4, and 10, respectively.
5. *Test your hypothesis* (using a statistical test): We will compare the overall average across farms (null hypothesis) with the individual averages for each farm. As usual, we will do this by computing variances and, in this case, using the *multiple-sample test for differences* as described next.

Multiple-Sample Test for Differences

Let's put into practice again these drawing skills that you are developing. We start by plotting in a graph paper the three values for each of the farms like this:

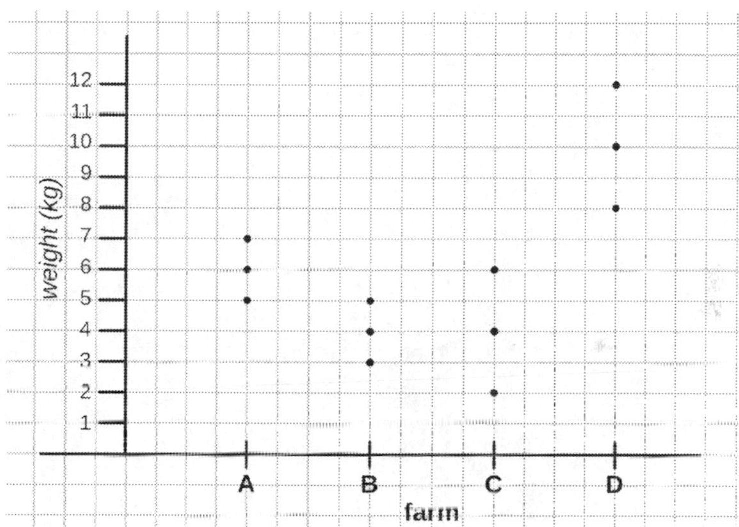

Like in the two-sample test, we draw a horizontal line representing the overall average (null hypothesis):

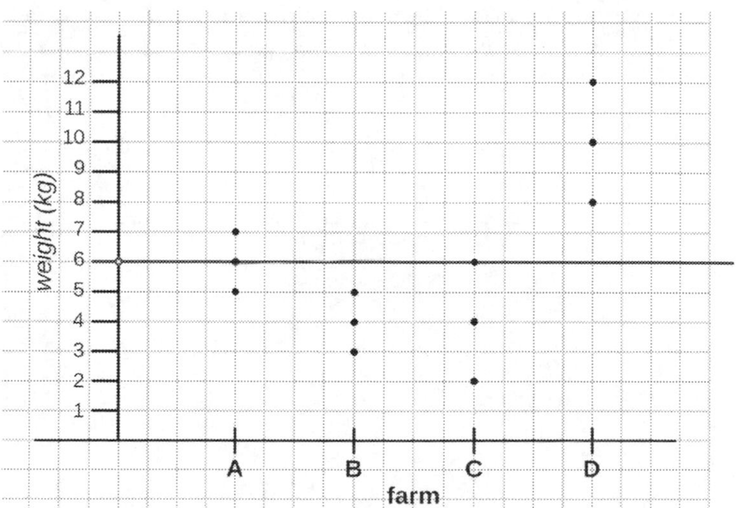

Next, as you did before, you can start computing the total sum of squares:

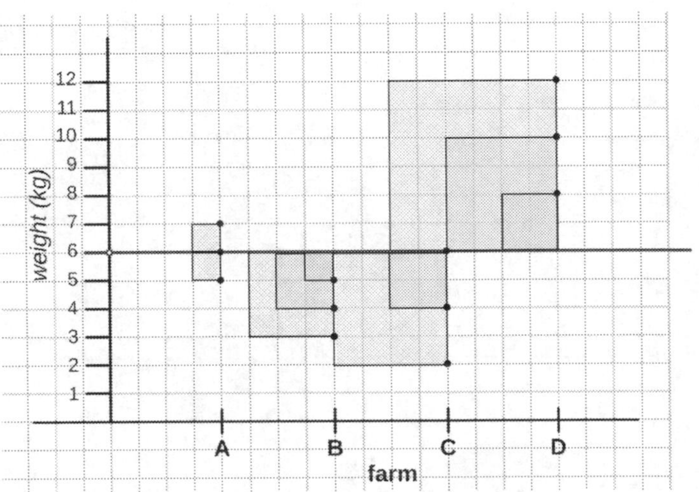

In this case, this is 92, associated with 11 degrees of freedom (remember, the total number of measurements minus one). We can start populating our ANOVA table already:

	SS	df	MS	F	*P*
MODEL					
ERROR					
TOTAL	92	11			

On a separate piece of graph paper, we plot again the 12 measurements, but this time highlighting the average in each farm:

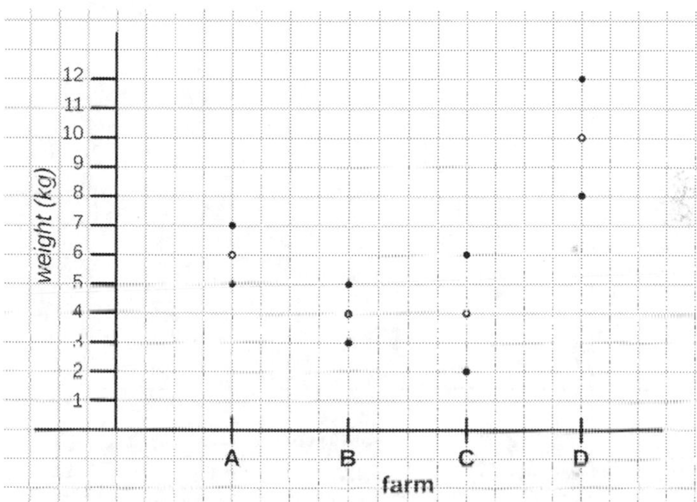

Then we join the average values with a line:

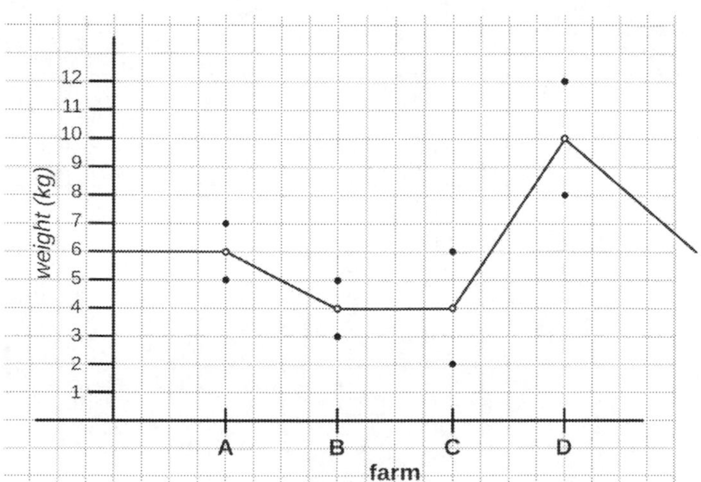

And finally, we can compute the error sum of squares:

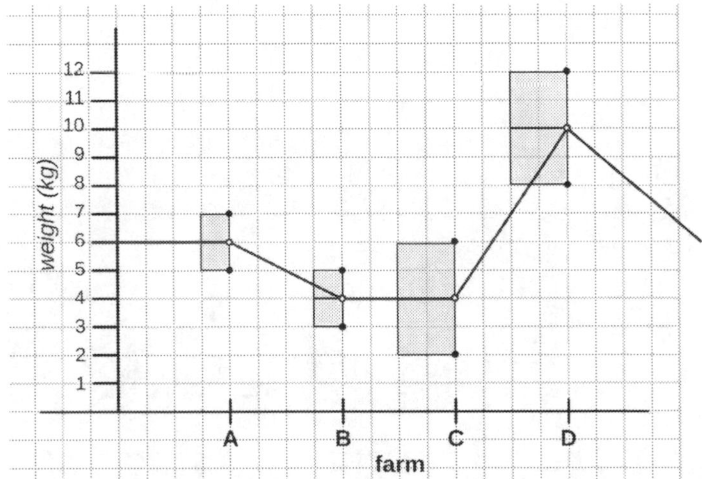

In this case, the error sum of squares is 20. The number of degrees of freedom is computed like in the two-sample test: the number of measurements minus the number of groups (farms in this case). In the ANOVA table, we have:

	SS	df	MS	F	*P*
MODEL					
ERROR	20	8	2.5		
TOTAL	92	11			

To find out the model SS, we can use the additivity property of sums of squares and of degrees of freedom. Or we can compare the two lines in our graphs like we did for other tests:

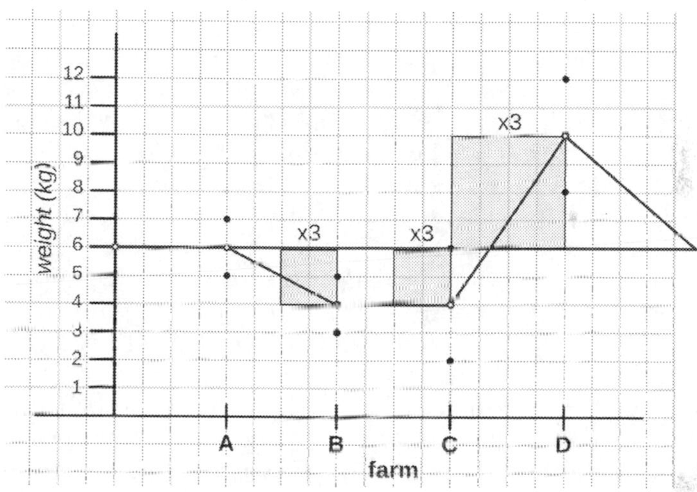

In any case, we can complete our ANOVA table and compute the *F* value:

	SS	df	MS	F	*P*
MODEL	72	3	24	9.6	0.004–0.006
ERROR	20	8	2.5		
TOTAL	92	11			

To compute the associated *P* value, we need to use a new table, as the number of degrees of freedom for the model is greater than 1. At the end of this book, you will find Statistical Table 4 for *F* distribution, *P* values with three degrees of freedom in the numerator. For other exercises, you will find useful the other available tables of *F* distributions for different degrees of freedom.

P Hacking: Statistics Done Wrong

At the beginning of this lesson, we talked about an alternative approach to do a multiple-sample test: perform all-against-all samples pairwise two-sample tests. As already discussed, one would need to correct for multiple testing (something we are not covering here). However, some people don't do that, yet they get away with it. How? By reporting only significant results. But this is statistics done the wrong way.

Imagine that a scientist (with not particularly good ethical behaviour) has ten drugs to test in mice if they increase the insulin levels in the blood. Our scientist performed ten two-sample tests comparing a control group with each one of the groups treated with one of the drugs. Of all the tests, only one had a low *P* value, specifically of 0.05. The scientist decides to publish only this result and ignore the other nine. When you, a curious researcher, read the paper reporting the results, you will be misled by interpreting that the probability of being wrong by rejecting the null hypothesis is 5%. But in reality, this *P* value does not take into account that nine other tests were performed.

This unethical behaviour during the report of results is called *P* hacking, and unfortunately is quite frequent in research, particularly that of clinical nature. There are two main reasons for it. First, scientists are often evaluated by the number and quality of publications, and sometimes this pressure to get significant results makes some scientists cherry-pick their results. Second (and more frequent), some inexperienced analysts are unaware of the importance of multiple testing and think this is OK. But it is not! And now you know it is not so, don't do it. Be honest.

TRY ON YOUR OWN

Alice doesn't trust Bob's analysis, simply because she believes that different mice belong to different strains, and therefore they may have a different metabolic rate. She identified three different types of mice that she identified as A,

B and C. The weight measurements (in grams) used in Bob's analysis can be classified by strain like this:

Breed A: 24, 28 and 20

Breed B: 30, 18, 21 and 17

Breed C: 27, 26 and 19

Is there any statistical evidence of the strain having an effect on measured weights? Perform a multiple-sample test for differences for these three groups.

Next

At this stage, I hope you have a clear idea of how statistical tests work in general. It's not rocket science and the principles behind these tests are relatively simple. Now you know how to compare one, two or more samples with the expectations of a null hypothesis to get some statistically relevant information. But, how about if some of these samples are related to other samples? For instance, if you want to investigate the effect of different petrol additives in car performance, but you are to test the effect in two different car models. Obviously, cars from the same model are related (same engine and more). What we need to do is to jointly estimate the effect of additives and models. The conceptual framework is the same as in previous tests, but you will need to push a bit your drawing skills.

Notes

1. In classic statistics, this approximation of the combined *P* value is known as the Bonferroni correction. It is very strict and other, less strict, approaches exist, but this is out of the scope of this introductory text.
2. In classic statistics, this is known as *one-way ANOVA test*.

10

Two-Way

Now that we have covered the basics of statistical testing, we are going to increase the level of complexity. Not that it is going to be very difficult, but we will need to build more sophisticated graphs. Here's where things start to get interesting, as we will now consider how two independent factors affect a specific outcome simultaneously with the use of two-way tests.

We are going to revisit the dataset from the previous lesson but consider a piece of information that we ignored. In particular, we noticed that farms A and B fed their chickens with wheat-based food, while farms C and D used corn-based food. However, farms A and C used spring water but farms B and D used filtered water in the food preparation. We have two factors: food and water. That is why this type of experiment is called a *factorial design*, and it's quite common in statistics. For a two-way test, we will proceed as follows:

1. *Set a hypothesis*: We hypothesize that the average weight of chickens in all four farms is the same, meaning that neither the food nor the water has any influence on chicken weights (null hypothesis).
2. *Decide on the statistical test based on your hypothesis:* We will perform a *two-way test for differences.*[1]
3. *Collect data (samples):* We take the weight in kilograms of three chickens from each farm. We use here the same measurements taken in the previous lesson, but we arrange the data in a two-way table:

		FOOD		
		wheat	**corn**	
WATER	**spring**	7, 5, 6	4, 5, 3	
	filtered	10, 12, 8	6, 4, 2	

DOI: 10.1201/9781003398820-10

4. *Estimate population parameters:* We need to estimate nine averages in this case. First, the average weight if all samples came from the same population as we did in the previous lesson. This total average is six. Then the average for each farm, as we did already: 6, 4, 4 and 10. In addition, we compute the average for each food and the average for each type of water. In other words, the average of the rows and columns in the two-way table. Indeed, we can add all the averages in the table for convenience:

		FOOD		
		wheat	**corn**	*average*
WATER	**spring**	7, 5, 6	4, 5, 3	5
	filtered	10, 12, 8	6, 4, 2	7
average		8	4	6

5. *Test your hypothesis* (using a statistical test): Unlike in previous tests, we need to perform several comparisons. We will compare the overall average across farms (null hypothesis) with the individual averages for each farm, but also with the averages of each row and column. Again, we will do this by computing variances, and in this case, using the two-way test for differences. In a later section, we will describe a unique feature of these types of tests: the interaction between factors.

Two-Way Test for Differences

To do this test the visual way, we are going to do a 3D plot (or something like it). We start by constructing a plot with a horizontal line (the so-called x-axis), another line going up (y-axis) and a third (new) line going towards us (z-axis). As in the previous lessons, the horizontal axis will have the values of the first variable (food): wheat and corn, and the vertical line the measurements (weights). The new axis will contain the values of the second variable (water): spring and filtered. It is important that we keep distances consistent when we work beyond the grid provided, for this, make a small paper ruler (in grey in the graph below) using the same graph paper:

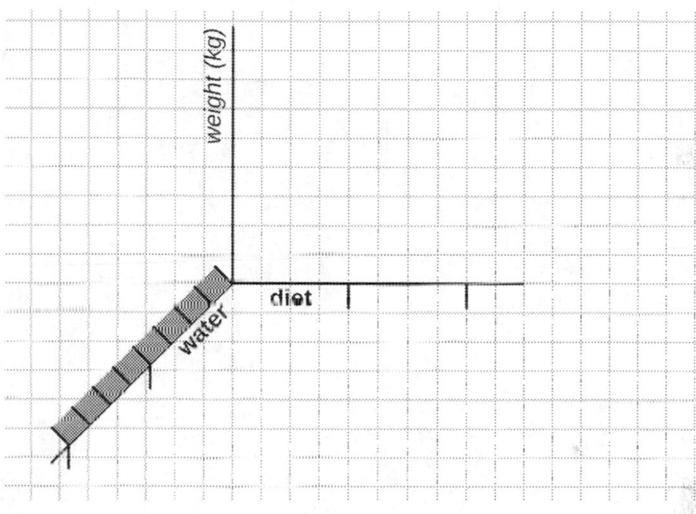

By drawing parallel lines, we will also find the intersection points across variables:

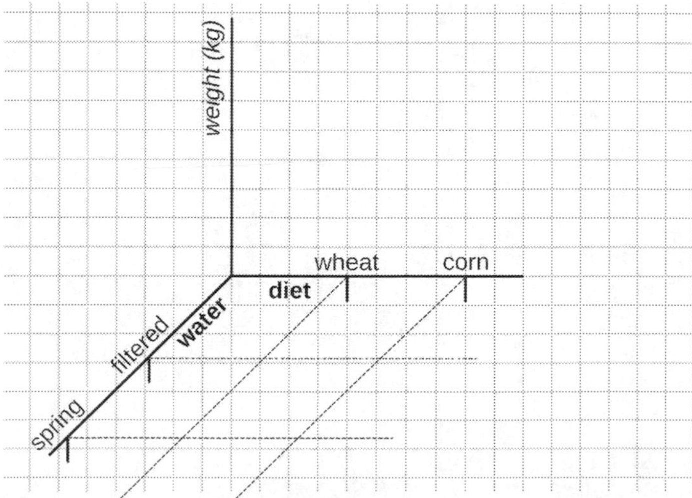

Each cell in our table will be represented by a vertical line from each of the intersections:

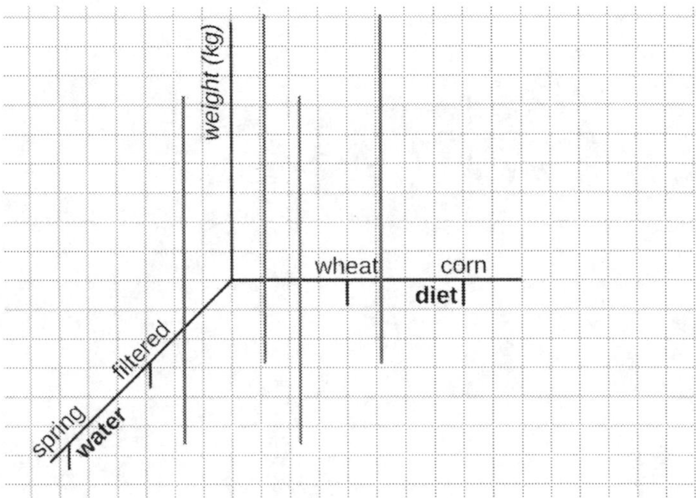

Using the paper ruler you made, you can start placing the sample values in the graph:

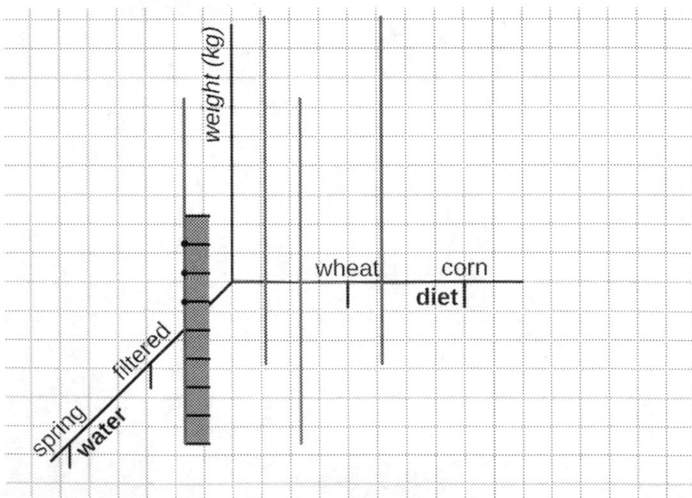

Your graph should look something like this:

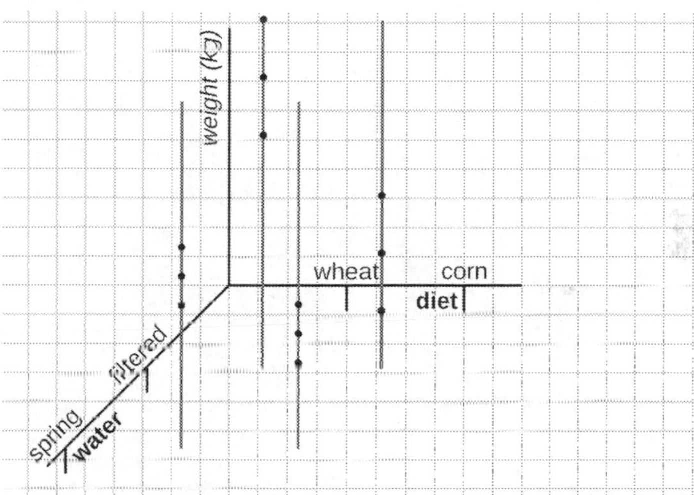

As in previous lessons, we start by finding out the total sum of squares. To do so, we start by plotting the total average in our graph. In our case, the total average across all samples is six, so you should identify the value 6 in each vertical line and join them with a plane:[2]

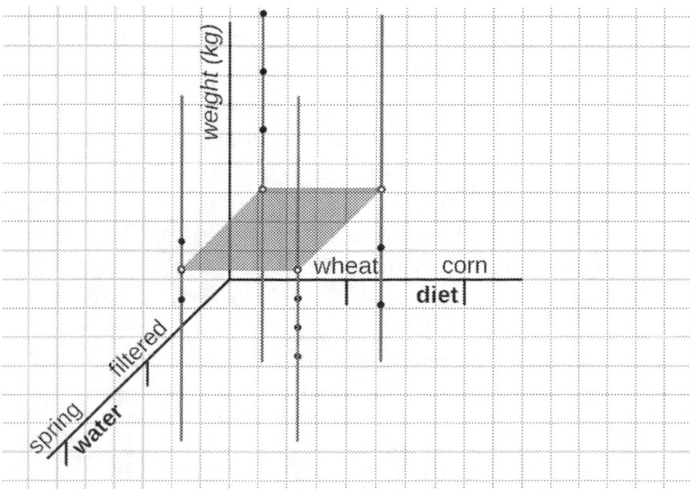

Now, we can start drawing squares connecting measurements and the total average plane:

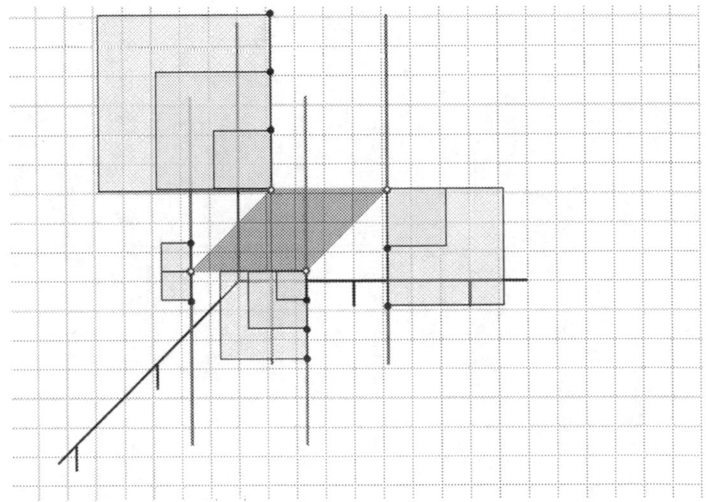

With the help again of our paper rule, you can find out the total sum of squares, so we can start populating our ANOVA table:

	SS	df	MS	F	*P*
MODEL					
ERROR					
TOTAL	92	11			

You may have notice that the two-way total sum of squares is identical to the multiple-sample total sum of squares form the previous lesson. This is not by chance. As we shall see, what we are going to do is to split this sum of squares into multiple components.

In the other tests, the error sum of squares was found out by counting the squares between the measurements average for each individual group, in this case, each cell in our table. The main difference is that now the lines connecting the averages form a pyramid:

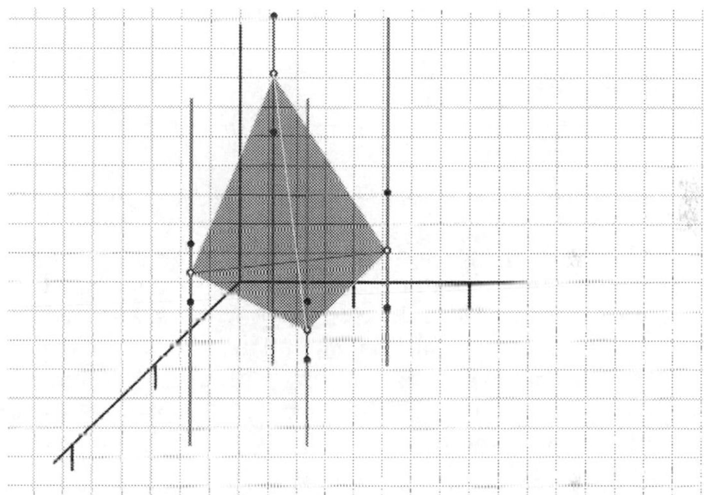

From which we can draw the connecting squares:

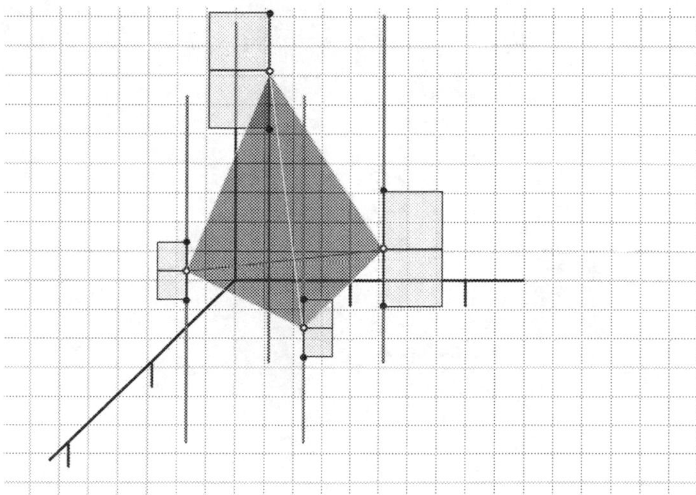

So, we can add the error sum of squares to our table. The number of degrees of freedom are the same as in the multiple-sample test so the table is the same so far:

	SS	df	MS	F	*P*
MODEL					
ERROR	20	8	2.5		
TOTAL	92	11			

Now it's when things get a bit more complex (yet more interesting). Since the sums of squares are additive, we could easily find out the model sum of squares. However, we have actually two models, one for food and one for water. We will need, therefore, to split the model sum of squares, starting with that of the model for food. This is given by the squares connecting the plane of the error with a plane connecting the partial averages for the two foods in our table:

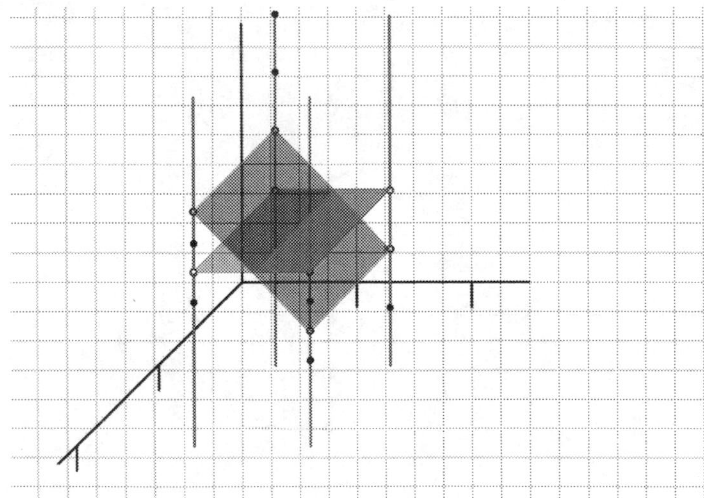

Remember that each square is repeated three times in this case, as we have three measurements in each group:

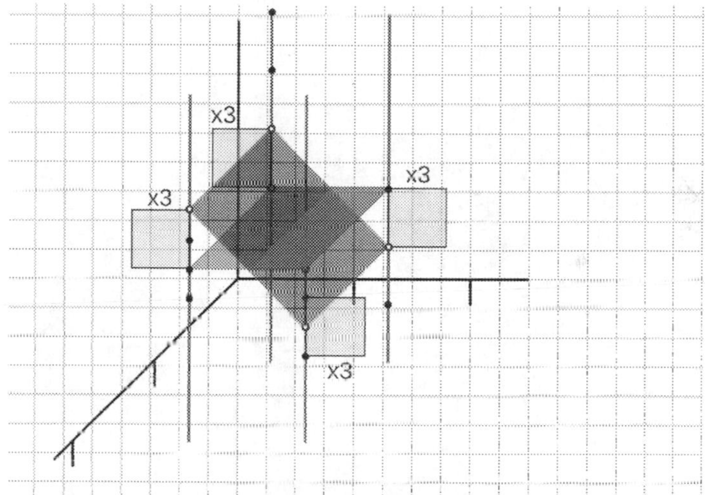

And we can add it to the table, and as in all previous tests, the number of degrees of freedom for this mode is the number of groups in the model minus one. In this case, we have two foods so just one degree of freedom:

	SS	df	MS	F	*P*
MODEL (food)	48	1	48		
MODEL (water)					
ERROR	20	8	2.5		
TOTAL	92	11			

Likewise, we can draw the plane for the water model:

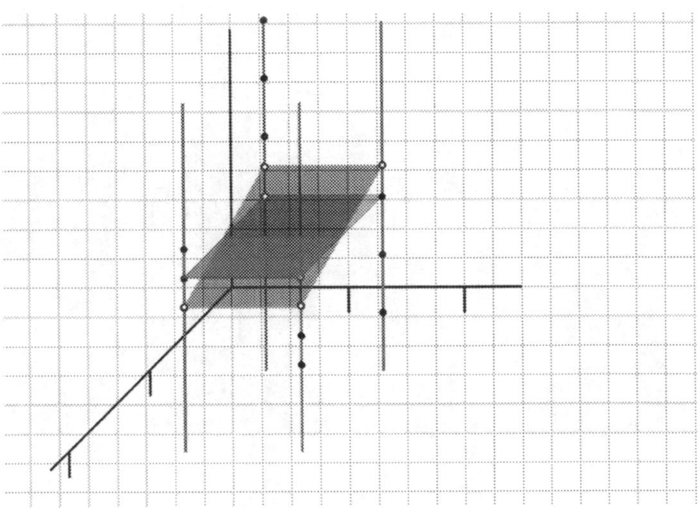

And count the squares:

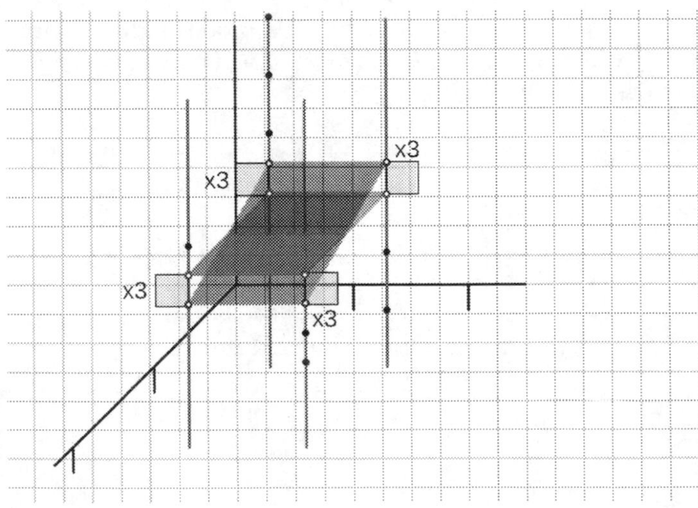

Completing the first part of the ANOVA table.

	SS	df	MS	F	*P*
MODEL (food)	48	1	48		
MODEL (water)	12	1	12		
ERROR	20	8	2.5		
TOTAL	92	11			

In the two-sample test, the model line was a rotation around the null-hypothesis horizontal line. Here, as we have two ways of comparing samples (hence the name two-way test), we have made two models by either rotating the horizontal plane in one or another way. Analogously to what we did in the other tests we can compute *F* and associated *P* values (Statistical Table 1) by comparing each model with the Error component:

	SS	df	MS	F	*P*
MODEL (food)	48	1	48	19.2	0.003
MODEL (water)	12	1	12	4.8	0.056–0.081
ERROR	20	8	2.5		
TOTAL	92	11			

From our ANOVA table, we conclude that there is a statistically significant effect of food on the weights of the chickens; however, the effect of water is not strongly supported. As you can see, a two-way test allows us to compare two models simultaneously.

But wait a minute! The sum of squares of the models and the error altogether is 80 (48 plus 12 plus 20), and this is different to the total sum of squares: 92. Where are the other missing 12 squares? Are we breaking the additivity rule here? Well, not quite, we are still missing one additional term.

Interactions

From the practical point of view, the test we have done could be understood as two independent tests, one for the effect of food and another for the effect of water. But there is an element we have not considered yet: the interaction between both effects.

Let's start with a simple example. There are many epidemiological studies investigating the impact of alcohol consumption and smoking on cancer incidence. Both alcohol intake and smoking have a positive influence on cancer development. Moreover, the joint effect in cancer development of alcohol intake and smoking altogether is higher than their separate effects. In other words, once you study the effects of alcohol and smoking separately, there is still a degree of cancer incidence that it is not explained by the two separate factors, but by the joint effect of both. This is called the interaction effect, and in our ANOVA table is represented as the interaction term.

Back to our farms, the interaction term will quantify the sum of squares attributed, not to the individual effects, but to the joint effect of water and food. If, for instance, filtered water has a component, not present in spring water, that modifies the absorbance of nutrients from corn, we will expect an interaction effect.

How to quantify interaction? We could actually draw graphs for the interaction term, but graphs will get messy and difficult to interpret. Instead, we will take advantage of something we have been repeating over and over again during this book: the additivity property of sums of squares and degrees of freedom. So, add an interaction row to your ANOVA table, and assign to it the missing squares and degrees of freedom to get to the total sum of squares, computing also the associate *F* and *P* values comparing the values with those of the Error row, like this:

	SS	df	MS	F	*P*
MODEL (food)	48	1	48	19.2	0.003
MODEL (water)	12	1	12	4.8	0.056–0.081
INTERACTION	12	1	12	4.8	0.056–0.081
ERROR	20	8	2.5		
TOTAL	92	11			

Now, our two-way ANOVA table is complete. In this example, only food has a detectable effect on the size of the chickens, while neither the type of water nor the interaction between water and food has a statistically supported effect.

TRY ON YOUR OWN

Bob understands Alice's complaint: the mice strain may have an influence on weight gain. He sorts out the data in a two-way table, but the number of measurements in some cells is too low. Hence, he expanded the number of experiments to make sure that there are three measurements in each cell. Here is the table:

		STRAIN		
		A	B	C
DIET	High in fat	24, 28, 23	30, 26, 22	27, 26, 25
	Low in fat	20, 21, 22	18, 21, 17	19, 20, 20

Perform a two-way test to evaluate the effect of the mice strain, the diet and the interaction between both factors.

Next

Up to here, you should be able to compare multiple samples independently or in a two-way form, taking into account an interaction term. This is good enough to perform many of the analysis you may need to do. But these tests can be generalized further to evaluate and quantify the association between samples. These tests are called regression tests, our next stop.

Notes

1. In classic statistics, this is known as *two-way ANOVA test*.
2. Plane as in a flat surface, not as in an aeroplane!

11

Regression

So far, we have been comparing two or more samples (or one sample and a given mean in the case of the one-sample test) to evaluate whether these samples come or not from the same population. But there is another family of tests to evaluate the association between variables. Among them, regression tests are not only the most used but also conceptually the most interesting (as we will see at the end of the lesson, all previous tests are actually instances of a regression test). A regression test evaluates whether a regression line is appropriate, being a regression line a model that allows us to make predictions.

We will start with an example. A narcotic is being tested in mice, and the investigators want to evaluate the effect of the dose on the hours of straight sleep after drug intake. Here's a table with the doses and hours of sleep:

Drug dose (milligrams)	Sleep (hours)
2	0
10	10
18	10
26	20

The variable fixed by the experimenter is called the *independent variable*, and the variable recorded after the experiment is called the *dependent variable*, as it depends on the value of the other

DOI: 10.1201/9781003398820-11

variable.[1] We can plot this relationship on the graph paper as a scatter plot:

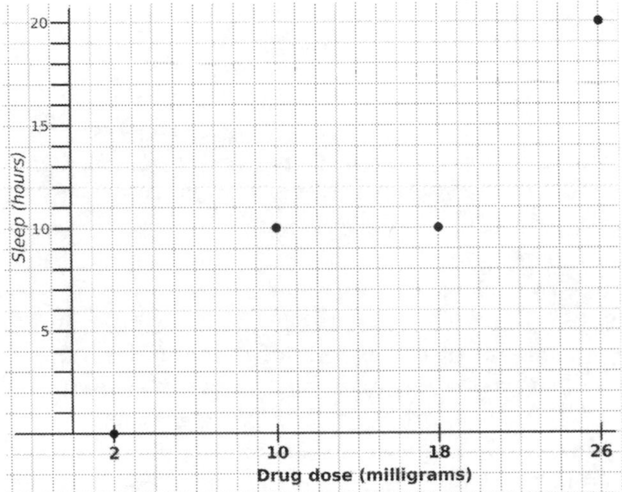

Ideally, we would have evaluated all possible drug doses, but as this is a continuous variable (recall from Lesson 1), it will be impossible to measure infinite values. Instead, we could fit a straight line that we could use, for instance, to predict the hours of sleep after taking 6 milligrams of the drug:

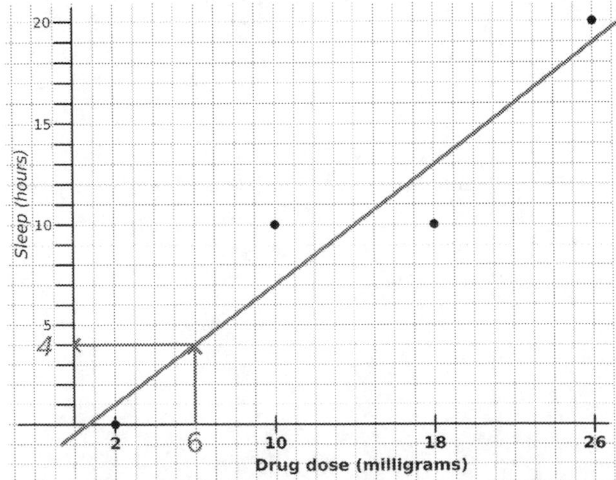

This line is called a *regression line*, and in this case, predicts that for 6 milligrams of drug, the mouse should sleep four hours. The goal of the *regression test* is to evaluate whether the fitted regression line is statistically different to the regression line expected under the null hypothesis: a horizontal line (in other words, that there is no association between the two samples). A regression test will be as follows:

1. *Set a hypothesis*: We hypothesize that the drug dose has no impact in the hours of sleep. In statistical terms, the hours of sleep and the drug dose are two independent and not associated populations (our null hypothesis).
2. *Decide on the statistical test based on your hypothesis:* we will perform a *regression test*.[2]
3. *Collect data (samples).* The investigator decides to use four drug doses: 2, 10, 18 and 26 milligrams (independent variable). The hours of sleep recorded for each dose were respectively 0, 10, 10 and 20 (dependent variable).
4. *Estimate population parameters:* Like in the previous tests, we eventually need two lines to be fitted: one for the null hypothesis and another for the proposed model. The former is the horizontal line given by the overall average of the dependent variable; in this case, it is six hours (this is equivalent to the total average in the previous tests). The latter is given by the least-squares regression line, as we shall see in the next section.
5. *Test your hypothesis* (using a statistical test): We will compare the regression line (model) with the null hypothesis line (average) and construct an ANOVA table exactly as in previous lessons. This is the regression test and will tell us whether the regression line is significantly associated to our data.

Least-Squares Fit of the Regression Line

If you recall from the two-sample test, we first counted the squares between the sample measurements and the horizontal line for the overall average, our null hypothesis. Then, we fitted a second line going through the total average and the average of each one of the two samples, and we called it the model line. You may also recall that this is the line that matched the least-squares criteria, that is, there is no other line for which the number of squares between the line and the sample values is smaller. For the regression line, we need to find the least-squares line as well, although it is not as straightforward as in the two samples case.

There is an equation that tells us exactly where the least-squares line is, but as I promised, there will be no equations in this book. In the rest of the lesson, you will be giving extra information about the regression line. In real-life statistics, a piece of software will find it for you so, no worries. But let's describe a few characteristics of the least-squares fit. First, like in the two-sample test, the line passes through the overall sample averages, but in this case, each average is in each of the two axes: 14 milligrams of drug (independent variable, horizontal axis) and ten hours of sleep (dependent variable, vertical axis). From a graphic point of view, it is like getting all the rotations of a line around that point and keeping that with the least squares count. Second, if the measurements are distributed 'appropriately', the line will pass near the averages of the 'left' and 'right' side points. In our example, if we split the independent values (horizontal axis) in two halves, the first two points to the left side have an average of six milligrams in the horizontal axis and of five hours in the vertical axis. Likewise, the second half (right side) has averages of 15 milligrams and 22 hours. Let's plot this line into the scatter plot:

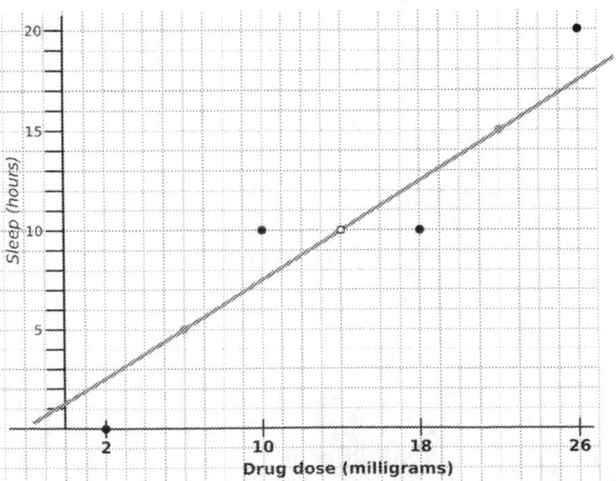

In this case, this line gets close to the least-squares, but it is not. If you play around with the rotation of the line, you will eventually find out that the least-squares line passes through the point defined by six milligrams/five hours. Let's plot it together with our first guess:

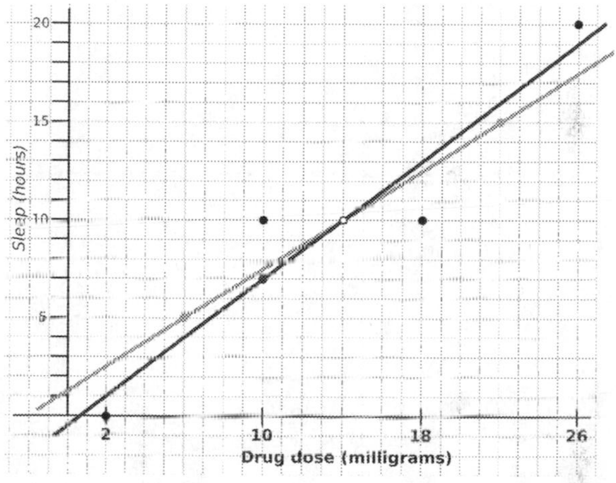

As I wrote earlier, you don't need to worry much since it will be either given to you (like in this lesson) or computed by a piece of software. Now we have what we need to do the regression test.

Regression Test

Since you have been doing this for a while, we can go quickly through the steps. Let's start by computing the total sum of squares. As in other tests, we add a horizontal line representing the average value of the variable represented in the vertical axis (our response variable), which is 10 hours, and connect the lines and the sample values with squares:

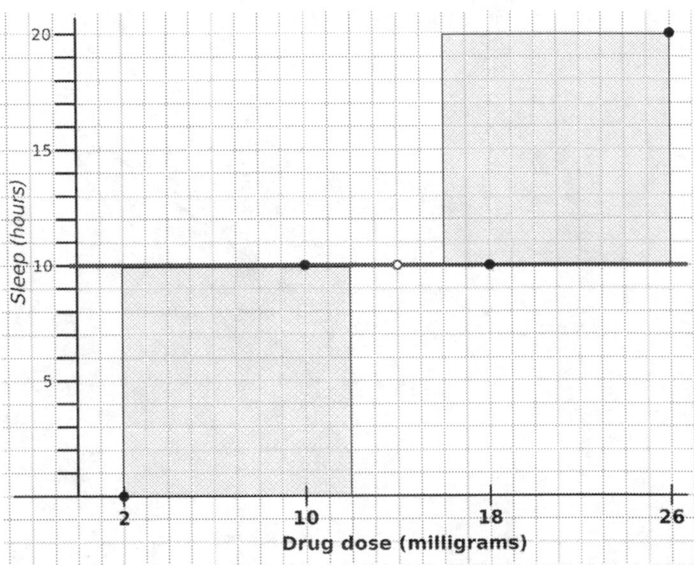

These are very big squares, and we count a total of 200 grid squares, which is our total sum of squares. Following the same rules as in previous tests, the number of degrees of freedom is the sample size minus one. We can now start populating our ANOVA table:

	SS	df	MS	F	*P*
MODEL					
ERROR					
TOTAL	200	3			

Next, we need to add our model line which is, indeed, the regression line. By counting the squares between the regression line and the actual data points we'll find the sum of squares of the error:

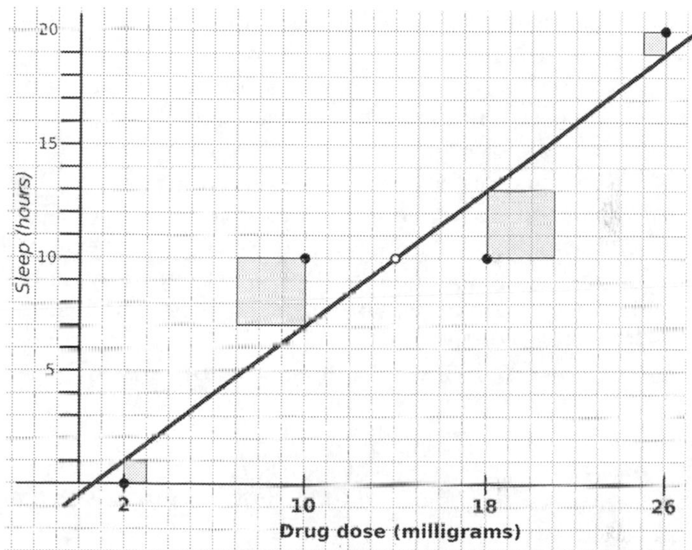

This in this case is 20. The differences between the regression line and the data points are also called in the statistics literature the *residuals*. Residuals are important in many other aspects of data analysis, so keep an eye on this concept if you decide to learn more statistics.

Now, to simplify things, let's use the additive property of the sums of squares and degrees of freedom to complete our ANOVA table. Since the model has one degree of freedom (as in the two-sample test), we can get our final table:[3]

	SS	df	MS	F	*P*
MODEL	180	1	180	18	0.051
ERROR	20	2	10		
TOTAL	200	3			

Even though there is a clear trend (regression) line that can be fitted, we don't have a strong statistical support, as our *P* value is relatively high.

Regression, Determination and Correlation

Regression is a method we have in statistics to predict outcomes, and the regression line is the tool used for this prediction, as we saw earlier in the lesson. The regression test tells us how much statistical support we have for the regression line, that is, whether the line is significantly different to what we would expect just by chance. But there is another useful measure, the *coefficient of determination*. The coefficient of determination tells us how well the regression line explains the data we have, and it is measured as the proportion of the model sum of squares with respect to the total sum of squares. In our example, the coefficient of determination is 90% (180 out of 200).

A common mistake made by an inexperienced analyst is to mix the concept of regression and *correlation*. Correlation is a measure of how two variables are associated, but it is not useful to make predictions, as regression is. The *correlation coefficient* is a number between −1 and 1 that quantifies how much two variables are associated (under a set of assumptions, as usual in statistics). Even though regression and correlation are different, they are mathematically related. Indeed, the coefficient of determination is exactly the square of the correlation coefficient between the two variables. Indeed, the coefficient of determination is usually called *R-squared*.

We will not talk more about correlation in this book as I promised no formulas will be presented, and we certainly need some formulas to understand properly the concept of correlation. But please, even though they are mathematically related, never mix up the two concepts: correlation and regression and two different things.

Linear Models and a Bit of History

If you read up to here, you may have the (correct) impression that I'm passionate about maths and about history. Yet, I'm doing a tremendous effort to not to show equations (the language of mathematics) or talk about history except in the introductory lesson. However, I cannot help myself to briefly discuss at this stage some of the history of the mathematical developments in statistical testing, as they have a lot to do with regression.

The concept of correlation (as described in the previous section) was developed in the 19th century as a way to associate the quantitative characteristics of parents and offspring in genetic studies. (Indeed, the development of modern statistics in general is tightly linked to the rise of genetics as a science in the late 19th and early 20th centuries.) But as I mentioned already, correlation can't be used for prediction, and animal and plant breeders were interested in ways of genetically select their stock to increase productivity, and human geneticists were interested in predicting the characteristics of children from that of their parents.[4] In this context, regression was developed, and became one of the most used statistical tools.

Already in the early 20th century, the first statistical tests to test differences between variables were developed. First the two-variable tests (so-called t-tests) and later on the multiple-variables tests (as we already saw), the two-way family of tests and other flavours of the ANOVA family of tests. The Eureka moment (well, more than Eureka was a progressive realization) came when statisticians started to connect regression and regression tests with other ANOVA-based tests. Like magic, it turned out that regression is a general framework under which to perform all of the other ANOVA tests (one or more samples, one or more ways). Actually, that allowed the mathematical exploration of more complex forms of ANOVA tests taking advantage of the mathematical properties of regression. The models generated in all these tests became unified by a set of common principles, mainly: that the data is normally distributed, and that the relationships are linear (as in 'straight lines'). These models are collectively called *linear models*.

Now that you know how to do a regression test, you can look back at all the analyses you already performed and you will easily see that all tests you have done are instances of a regression test where the groups to be compared constitute the independent variable and the measurements correspond to the dependent variable. You have been building linear models all along. Indeed, this book is only possible thanks to the existing unifying principles to all classical statistical tests: regression lines and sums of squares. It is thanks to that that we can do statistical tests by using a ruler and graph paper. Now you know the secret. But don't keep it to yourself, spread the word!

TRY ON YOUR OWN

Alice is now convinced that a poor diet is behind the overweight of her mice. But she wants to evaluate whether a good exercise programme could help mice to gain control back over their weights. To do so, she devised a ten-minute exercise routine for mice. Then, she gathered ten mice, all weighing 28 grams. Then she kept each mouse under equal conditions for two weeks, except for the number of exercise routines that each mouse does per day. These are the final weights for each mouse under a different number of exercise routines per day:

Number of exercise routines per day	Weight after the experiment (grams)
0	28
1	26
2	27
3	24
4	25
6	21
7	22
8	19
9	20
10	18

Perform a regression test to evaluate whether the amount of exercise has an effect on the final weight of mice. (Hint: The regression line passes through the point: 0 exercise, 28 grams.)

Next

Up to here, we have covered the so-called classical linear models, and most statistical tests in the scientific literature are one of these tests or a variation. But we have to keep in mind, before we apply any of these tests, that we made a few assumptions. Mainly, we assumed that the populations from which the samples were drawn are normally distributed. But that is not always the case. What can we do, then? In the next lesson, we will see how to deal with non-normal populations.

Notes

1. In other contexts, these are also called the *explanatory* variable (independent) and the *response* variable (dependent).
2. In classic statistics, this is known as *regression test*.
3. If you struggle to fill this table at this point, you may revise the previous lessons to make sure you understand how MS, F and P values are computed.
4. Some of these studies were done in the context of *eugenics*. Not everything in the history of statistics is something to be proud of, I'm afraid.

12

What If My Data Is Not 'Normal'?

All the tests we have done so far assume that populations are distributed normally. In Lesson 3, we saw that natural populations are often (actually more than often) normally distributed. That means that if we take a large sample and plot a histogram, this will look bell-shaped with a peak around the mean value. But what if the populations are not normally distributed?

For this example, we are going to imagine a flock of pigeons released at some distance from their pigeon holes at the same time. An observer is at the pigeonholes' room timing when the pigeons arrive. She observed that after two minutes, four pigeons arrived. After four minutes, another seven made it to the room. After eight minutes, eight more pigeons were there, and she has to wait up to 16 minutes for another five pigeons to make it to the pigeon-hole room. Finally, after 32 minutes, the last pigeon arrives. Let's build a histogram, using a bin size of five (see Lesson 1):

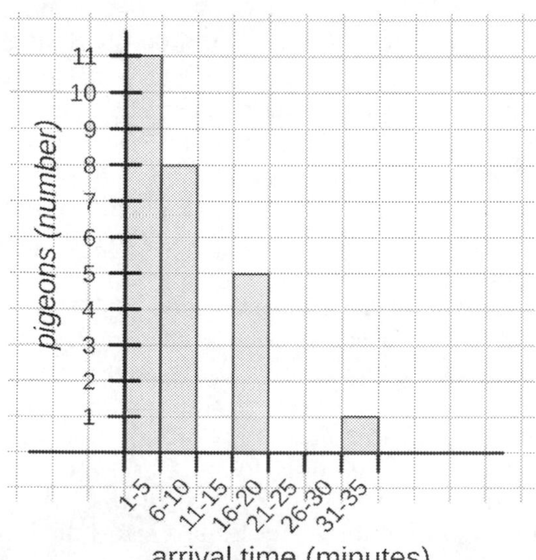

DOI: 10.1201/9781003398820-12

As you can see, the distribution does not look much like a normal distribution. If we were to compare this sample with another sample, the assumption of normality of the tests we described is violated. However, we still have a few options. We will be discussing two of them here: transforming the data, or using another family of tests often called non-parametric tests. But before that, let me explain why in nature almost everything looks normal, but sometimes it doesn't.

Why (Almost) Everything Is Normally Distributed

A fundamental theorem[1] in statistics is the *central limit theorem* (CLT). It states that if you sample from a population, no matter how the population values are distributed, the average values of multiple samples will be normally distributed. It is an asymptotic property of sample averages: that is, averages get closer to a normal distribution as the sample size gets larger. This is related with the law of big numbers that we discussed in a previous lesson. The CLT is a mathematical theorem, not an empirical observation, and it has profound implications in many fields in science, mathematics and beyond. But from the point of view of nature itself, it may explain why almost we see is normally distributed.

Think about the weight of a person. This is determined by the influence of many factors: multiple genes and their interactions, current and past eating habits and more. The current weight of a person is the average of the contribution of these factors (actually, a weighted average as different factors will contribute differently). Therefore, by the CLT, it is expected that weights are also distributed normally.

Some measurements from nature, however, are not normally distributed. Why? This could be due to multiple reasons. For instance, the character we are measuring may depend only on a limited number of factors so the CLT does not hold, or perhaps the weight of one of the factors masks any other influencing factor, or the quantity we measure simply follows a different distribution. For instance, the waiting time of an event (like in the example before) follows a different distribution (the so-called exponential distribution).

Transformation

Sometimes, measurements from a sample that are not normally distributed would be normally distributed if the measurements were transformed in a specific way. For instance, the example at the beginning of the lesson can be 'made more normal' with the use of logarithms.

To understand what a *logarithm* is (in case you don't know this already), we need to explain first what an *exponential* is. Imagine that you have a single bacterium that grows and split into two after one minute, so you have now two bacteria. After another minute, you have each bacterium producing two bacteria, so you will have four. Every minute you double up the number of bacteria. So, after three minutes you have 8, after four minutes there are 16 and so forth. After 20 minutes, we will have more than a million bacteria, and within an hour, the number will reach (at least theoretically) a billion of a billion bacteria. This is called exponential growth and is the reason why pathogens spread so quickly in a population. The logarithm is the reverse of the exponential. For instance, in this case, if we define a logarithm in base 2 (because we are doubling up every minute), the logarithm of the number of bacteria will be the minutes from the start: that is, the logarithm in base 2 of two is one, and the logarithm in base 2 of 16 is 4. You can build a table like this:

The logarithm in base 2 of	is
2	1
4	2
8	3
16	4
32	5
64	6
128	7

Importantly, logarithms can be defined in other bases. For instance, if you multiply by ten every time, the reverse of the exponential 10 function will be the logarithm in base 10 function. Here is a table:

The logarithm in base 10 of	is
10	1
100	2
1,000	3
10,000	4
100,000	5
1,000,000	6
10,000,000	7

Now take the data from the example at the beginning and transform with a logarithm in base 2 (or $\log_2$ in short) the waiting times. If you plot a histogram of the new dataset, it will look like that:

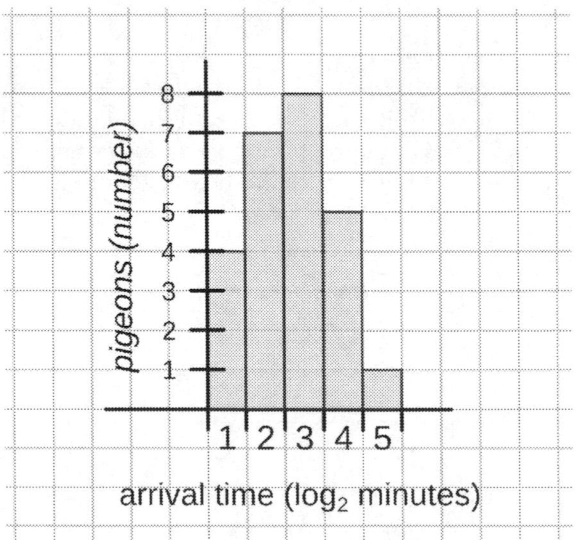

That's pretty normal, isn't it? This data is ready to be compared with other samples using the tests you already know.

This is just a glimpse of transformation techniques as there are myriads of ways of transforming the data (logarithmic transformation is just one of them). Data transformation is a bit of an art, and it is not always free from technical problems. The advice I give to my students is, if you are not sure why and how are you transforming your data, don't do it.

Non-parametric Tests

If you want to compare two (or more) samples whose underlying populations are not normally distributed, the standard tests we've seen may not work. When you compute the sums of squares and, finally, your F value, the P value you find is based on the assumption that all populations are normally distributed. During these computations, the different means and variances are estimated following the assumption of normality. These values are parameters of a normal distribution, and that's why the tests we have seen so far are often called parametric tests.

Alternatively, you can generate custom distributions from your samples, without making any assumptions. In this sense, you will not need to compute the parameters of a given distribution: your sums of squares will be distributed according to your generated custom distribution. These tests are called collectively non-parametric tests. In reality, there are parameters being computed in a non-parametric test, but this is a bit technical and, since statisticians have adopted this nomenclature, we will adopt it here as well.[2]

Typically, a non-parametric test is based on ranks: that is, the order of the values rather than the values themselves. Once we use ranks instead of values, for large samples, we can assume that these ranks are (roughly, in a very relaxed way) normally distributed, so we can perform the tests we now over ranks. This is a big simplification, and the non-parametric tests implemented in standard statistics computer programs use more sophisticated and realistic distributions. But to understand how these tests work, it is a reasonable first approximation.

For this test, we are going to investigate whether a new fertilizer has an impact on production. A usual old fertilizer is utilized in four fields, and a new (supposedly improved) fertilizer is used in other four fields of similar characteristics. The first four fields produced 1, 5, 11 and 13 tons of vegetables, respectively, and the four fields with the new fertilizer produced 4, 10, 12 and 22 tons, respectively.

Instead of working directly with the values, we sort them all and rank them, from 1 (the smallest) to 8 (the largest). In the following table we can see the values for both groups and their corresponding ranks:

	Sample value (tons)	Rank
Old fertilizer	1	1
	5	3
	11	5
	13	7
New fertilizer	4	2
	10	4
	12	6
	22	8

Now, we perform a two-sample test as we did before, but this time using the ranks instead of the original sample values. Let's start this time by finding out the sum of squares of the error:

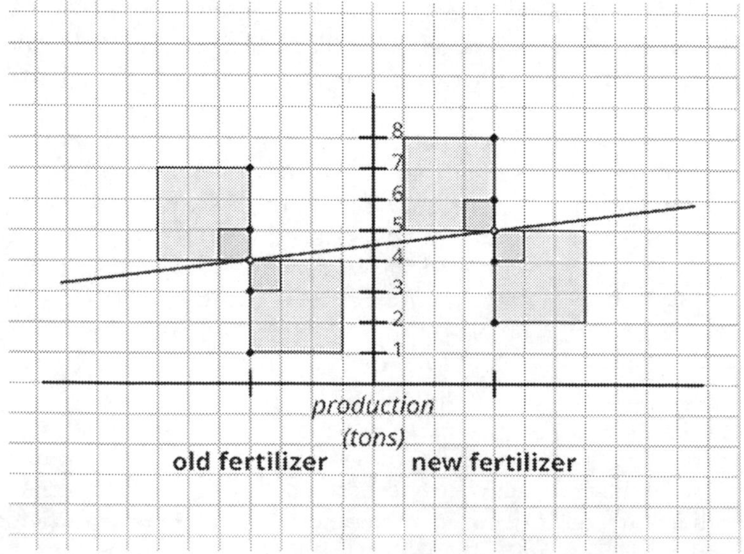

From the graph, we compute a sum of squares of 40. To make things simpler, and take advantage of the additivity property of squares, we are going to compute next the sum of squares of the model. To do so, we first calculate the overall rank average, which is four and a half:

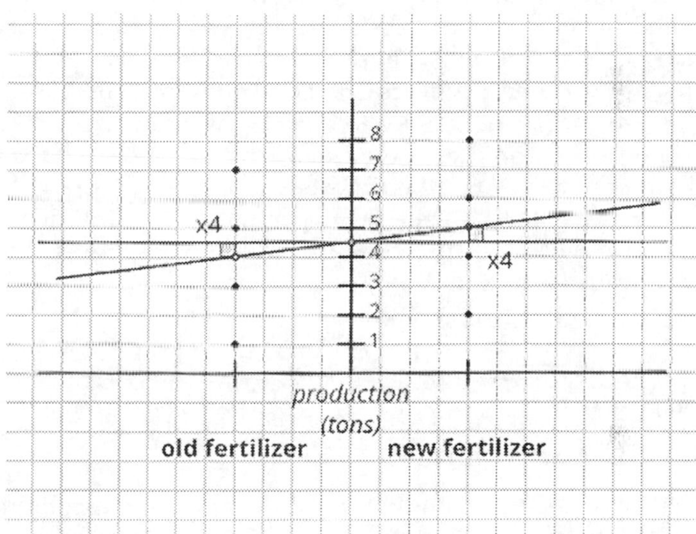

As you can see, we have squares that are a fraction of the grid squares. Each square is actually worth one-quarter of a full square. Since we have eight of them, we conclude that the sum of squares of the model is 2. The reason we compute this first instead of the total sum of squares is to avoid dealing with several squares of fractional sizes. But that's OK as, like before, we will rely on the additivity property of the sums of squares and the degrees of freedom. Using the same rules as in the standard two-sample test, we can get our ANOVA table ready:

	SS	df	MS	F	*P*
MODEL	2	1	2	0.3	>0.356
ERROR	40	6	6.7		
TOTAL	42	7			

From this test, we concluded that the observed difference is not statistically significant.

There exist non-parametric tests analogous to all the tests already seen (and more). Here we briefly described how to do a non-parametric two-sample test,[3] but you can do non-parametric tests for paired samples,[4] multiple samples,[5] two-way comparisons[6] and even regression, correlation and beyond. Non-parametric statistics is a topic on its own.

Why not use non-parametric tests all the time? Because we lose power, that is, the ability to detect real differences. This is why non-parametric tests are used when you can't use a standard parametric test. So, the first step is to have a look at your data. If it looks 'normal', perform a parametric test. If it doesn't, non-parametric is the safest choice. This seemingly trivial choice is often a cause of argument between scientists!

Not All Is About Normality

This lesson is to show you that when the normality assumption is violated, there are still ways around. However, things are a bit more complicated in real life. How do we know if the data is normally distributed or not? To know that, there exist a number of statistical tests, called *normality tests*. Yes, in statistics, we often perform a test to know which test we can perform.

There are also other assumptions of the tests we described that may be violated. For instance, the tests in this book assume that the variances of different samples are equal. Actually, this is a 'soft' assumption as tests are often robust to this violation. But in some contexts, it needs to be taken into account. Another assumption is that, if we compare two or more samples, these have to be independent. In Lesson 7, we introduced the paired-sample test as a way to deal with independence, but sometimes the dependency of samples is quite complex. A last assumption to be considered here is the sample size: standard tests will not work if the sample sizes are small. I will devote a whole section about it in Lesson 14.

TRY ON YOUR OWN

Alice is challenging Bob's two-way as she believes the data is not normally distributed. Hence, Bob is determined to compare again the two original samples but with a non-parametric test, to convince Alice that the results will still hold even if we don't assume normality. Perform a non-parametric test of the data you already analysed in Lesson 7.

Next

We are getting to an end. We covered all the tests I wanted to explain about comparing continuous samples. But we left behind an equally important type of variable: discrete variables, that is, count data. Before we finish our statistical journey, we still have to learn how to deal with discrete rather than continuous samples. In the next lesson, we will learn how to count (statistically speaking).

Notes

1. In mathematics, a theorem is an idea, an statement or a relationship between numbers that has been proved.
2. In statistics, we often refer to *permutation tests* when we estimate the underlying distribution of a population by randomly shuffling sample values. This seemingly naive approach is actually quite efficient to compute exact *P* values without assuming normality. Indeed, some statisticians use the category of *exact tests* for those tests that compute exact *P* values using randomly generated distributions or, in a few cases, well-defined probability functions.
3. In classic statistics, this is known as the *Mann-Whitney U test*, or sometimes the *Wilkoxon rank-sum test*.
4. Known as the *Wilkoxon signed-rank test*.
5. *Kluskal-Wallis test*.
6. *Scheirer–Ray–Hare test*.

13

Counting

Up to here, all tests we performed were on continuous data (lengths, weights, time etc.). However, there is a type of data that we have been ignoring so far: discrete variables. That is, the natural numbers we use for counting (0, 1, 2, 3 . . .).[1] When we use discrete values to count anything, we talk about *frequencies*. For instance, if you want to compare the number of eggs between multiple nests, or the number of patients surviving after a specific treatment, you can't use the tests we described so far. To deal with frequencies, we need a new approach. But don't worry, you already know the basics and, unsurprisingly, these tests are also based on finding out squares.

We are going to explore whether a representation of students reflects the national composition of the class. Let's suppose that we know that nine in 12 students are British; two in 12 are Europeans[2] and only one out of 12 is from elsewhere. So, we open 12 positions as students' representatives. Are the 12 students chosen a good representation of the national composition of the class? The key is to compare the *expected* frequencies with the *observed* (actual) frequencies using a specific test. For a test of differences in frequencies, we proceed as follows:

1. *Set a hypothesis*: We hypothesize that the observed number of students is no different to the number of expected students per national category. In other words, both sets of counts (expected and observed and treatment) come from the same population (this is our null hypothesis).
2. *Decide on the statistical test based on your hypothesis:* We will perform a *test for differences in frequencies.*[3]
3. *Collect data (samples):* We observed that in our representatives panel, we have six British students, four European and two from somewhere else.

DOI: 10.1201/9781003398820-13

4. *Estimate population parameters:* The population is defined by the expected values. In our case, the expected values are 9, 2 and 1 for British, Europeans and Elsewhere categories.
5. *Test your hypothesis* (using a statistical test): Using the test described later, we will compare the expected counts (null hypothesis) with the observed counts of students per nationality group.

The concept of *goodness-of-fit* is also defined in this context to refer to how good the observed counts fit to the expected counts.

Test for Differences in Frequencies

On a piece of graph paper, draw a plot like those you did before, but indicating the observed values, connected by lines:

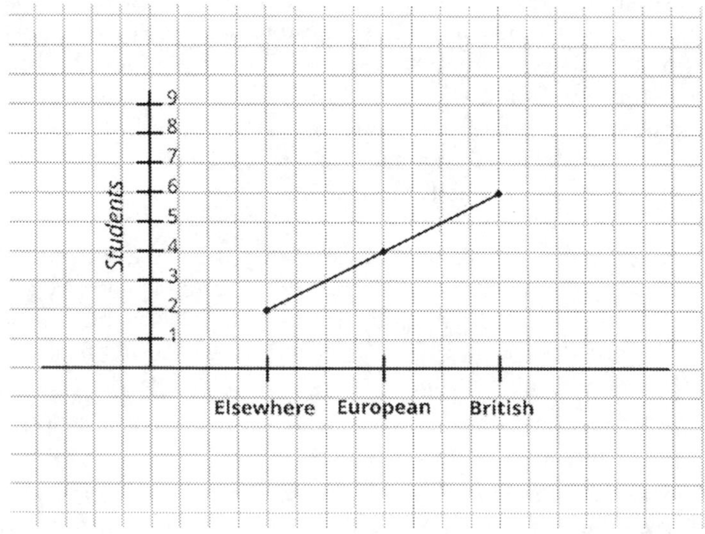

In a different colour, plot the expected values in the same graph:

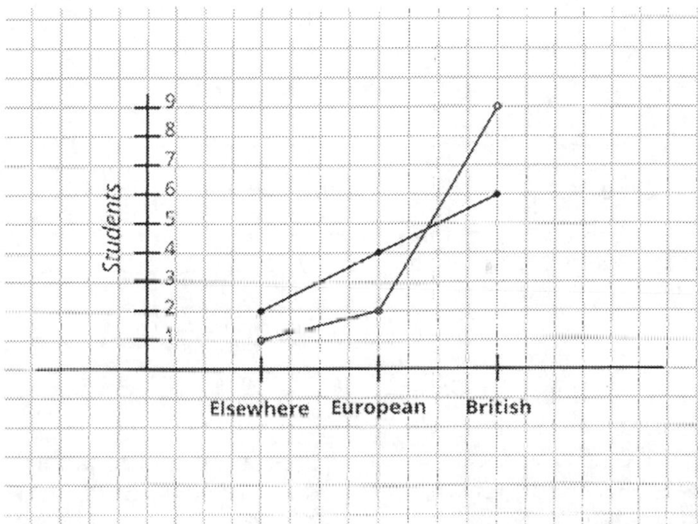

Like in previous tests, connect with squares the observed and expected values for each category:

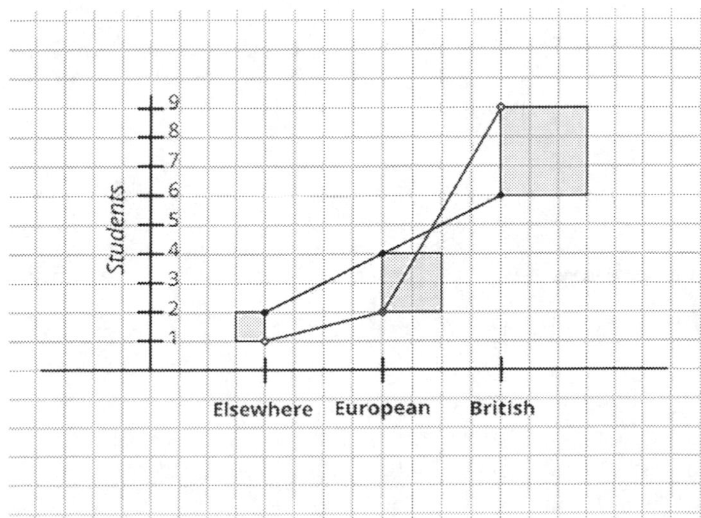

But now is when this test differs from the others we made already. For each of the categories, we count the number of small squares (the *squared difference*) and then we divide it by the expected value (we call these quantities the *standardized squared differences*. In the following table, we do these calculations:

Category	Observed	Expected	Squared Difference	Standardized Squared Difference
British	6	9	9	1
European	4	2	4	2
Elsewhere	2	1	1	1

The total of adding all the standardized squared differences is the statistic X^2, which in our case is 4.

How is the distribution of X^2 values if the null hypothesis is true? In this case, this value is expected to be distributed as a *chi-squared distribution* with a number of degrees of freedom equal to the number of categories minus one, which in our case is 2. We already discussed this distribution in Lesson 3 and, in fact, the *F* distribution is made up of two chi-squared distributions so both types of tests are somehow related (as we shall see in the next section).

By looking at Statistical Table 5 at the end of the book, a X^2 value of 4 with two degrees of freedom gives us a *P* value of 0.135. We conclude that the differences between the observed and expected values are not significant.

Beyond the Linear Model Paradigm

Except for the test described in this lesson, all other tests were instances of linear models (as we already explained in the lesson about regression). How it comes that the test described in this lesson is not a linear model? After all, it is also based on lines and squares, although in a different way.

Linear models assume that the populations from which the samples are derived are distributed normally. To be more precise, these tests assume that the differences between the model and the data are normally distributed. But in the case of discrete data, this is not the case, simply because a normal distribution is continuous and discrete data is, unsurprisingly, discrete. Is it possible to build a model, analogous to a linear model, but assuming a different distribution for our samples? Yes indeed, and these types of models are called Generalized Linear Models.

A *Generalized Linear Model*, or GLM, is a generalization of a linear model such that the data can be distributed other than normally. Even more, if for instance in a regression the association between variables does not follow a straight line, the GLM can be modified such that other than straight lines are considered. Hence, a GLM is defined with two components: a distribution component and a relationship component.

For instance, if you define a GLM with a normal distribution and a straight-line relationship, what you have is a linear model like those we have seen throughout the book. If you define a GLM with a Poisson distribution[4] and a logarithm relationship, what you have is the test we described in this lesson.

In summary, most of the statistical tests you will ever encounter are instances of a GLM. This is a relatively recent development in mathematical statistics, and its formulas can be quite intimidating for a stats newbie, yet the concept is simple and easy to grasp (and I hope I described it clearly).

TRY ON YOUR OWN

Alice, now convinced that she needs to provide a healthier diet to her mice, prepared three bowls with healthy food from three different supermarkets, to see if the mice have any preference. Out of 30 mice, she observed that 20 preferred the food from Trisco supermarket, 5 preferred that of Aldo store and the other 5 from Wait-flower. Are the observed counts different to the random expectations?

Next

We are done with tests and analyses. Well done! In the next and last lesson, I will give you a brief overview of more advanced topics that you should take into account if you decide to learn further statistics.

Notes

1. Interestingly, the number 0 is not always included in the list of natural numbers as it is a special number on its own. But let's not be so picky. For us, 0 is as natural as the others.
2. Curiously, in Britain, people call Europeans to all non-British Europeans, even though Britain is a part of Europe.
3. In classic statistics, this is known as a *chi-squared test for differences*, a *chi-squared test for goodness of fit* or simply a *chi-squared test*, but please be aware that there exist other (different) types of chi-squared tests.
4. This is a distribution that was developed to study *rates*, as in the number of events happening per period of time.

14

Size, Power and Effects

The goal of this book has been always the introduction of statistical analysis without the use of any formula. If you understood the basics of statistical testing, the necessity of a well-formulated hypothesis, why we use sums of squares and the various ways we can compare variances to test the hypothesis, this goal has been successful (and I hope that's the case). However, this book is about principles more than practical skills. If you were to apply the methods here described with large and real samples, the computation of the sums of squares will get very tricky. In this case, you will need formulas or, most likely, computer programs. But the principles underlying these formulas/programs are the same to those you already learned here. Thus, you will be able to understand what for others would be a black box. It is important to keep this in mind as you read this final lesson, which does not contain exercises or examples, but a discussion about the limits and the actual challenges of statistical analysis. We will discuss in turn some of the main points you should consider before you start applying your newly acquired skills.

Effect Size

Statistical testing has allowed us to discriminate perceived differences due to random variation from true differences due to a condition. But true differences supported by a statistical test do not necessarily imply that the difference is meaningful. For instance, you may be testing a drug to induce sleep in humans, and after some trials, you conclude that the drug has an effect in the sleep time, and that this effect is significant with a very small P value. However, you also notice that the difference between the averages is of only two minutes of extra sleeping time after drug intake.

DOI: 10.1201/9781003398820-14

The observed difference between the sample means is called the *effect size* and, in this case, it is almost negligible.

This is why it is important to report, not only *P* values (and the complete ANOVA table) but also the means for each group (effect) and, if possible, confidence intervals. In statistics, in general, we want to know whether a difference is significant and also how large (or small) is this difference.

Sample Size and Power Analysis

The main assumption of the tests we learned is that the sample measurements are normally distributed. As we saw already, this is a consequence of the central limit theorem, which in turn depends on the law of large numbers. But the sample sizes in the examples in this book are not particularly 'large', as in most cases, we worked with a sample size of three or four. Any statistician will criticize this. But as I wrote already, this book is about principles. Once you understand the principles, you should be able to understand how tests work with real sample sizes.

But how large should the sample size be? To estimate the required sample size, we should do something called *power analysis*, and it is usually the first step before doing any sampling. A power analysis is based on the required significance level (*P* value) and consists of reverting the statistical test to estimate the sample for which a difference, if existing, will be detected by the test. Power analysis takes into account not only a significant level but also an expected effect size. This type of analysis is beyond this introductory text as it requires some maths that we did not cover. But if you are learning more stats, power analysis is one of the methods you will need.

What if I still have small sample sizes? The first option is to perform a non-parametric test (see Lesson 12). Alternatively, one can use something that we did not cover here: exact tests. These are based on exact *P* value computations using specific probability functions, or the computation of exact *P* values based on distributions that we generate using the very same samples (we briefly touched on that in Lesson 3). But as a rule of thumb: if you have small sample sizes, try and get more measurements!

Multiple Testing

Social media success is (partly) based on the concept of *confirmation bias*. As you follow people with similar interests and even ideology, it is common to find news and memes that agree with your own views, giving the false impression that people not thinking like you are wrong and oblivious to the available evidence. Believe it or not, this also happens in statistical analysis, and happened way before the development of social networks.

Imagine that you are in your lab, testing the effect of a given drug to induce sleep and find that the effect is not significant (large *P* value). The next day you try another drug, and so forth, until the tenth attempt, in which you find a drug with a noticeable effect and an associated value of 0.04. Happy with your results, you publish your findings in a journal and report that *P* value. Indeed, the probability of wrongly rejecting the null hypothesis (as we already saw) is about 4%. However, this *P* value is not valid if you perform multiple tests. This is a type of confirmation bias or, in the context of statistics, *P* hacking (something we also discussed in Lesson 9).

An honest statistician should perform something called *multiple testing correction*. That is, it should take into account not only all the tests performed but also all the possible test that could be done. For instance, if you want to measure the joint effect of two drugs between a set of four drugs, you should take into account that you can make six different pairwise comparisons (even if you don't perform all of the comparisons). There are several methods, being the *Bonferroni correction* the most conservative. In this method, the *P* value threshold is divided by the number of comparisons. For instance, if you aim to reject the null hypothesis for a *P* value of 0.05 in a test involving 10 comparisons, the equivalent threshold will be 0.005.

Bonferroni is a very strict procedure and often we use less strict (yet justified) methods to control the probability of rejecting at least one true null hypothesis. Alternatively, we can allow to have some wrongly rejected null hypothesis, as long as we control how many of those we have. In this case, we can use something called *False Discovery Rate (FDR)*, in which we specify how many (on average) wrongly rejected null hypothesis (false positives) we will have among the multiple tests that we are performing.

The maths behind multiple test corrections is beyond the aims of this book, but it is something to take into account if you perform statistical analysis, particularly nowadays where we make myriads of tests in routine data analysis procedures.

Fixed, Random and Mixed Effects

I cannot finish this introductory text without at least introducing one of the current hot topics in statistical practice. When we compared two or more samples, we fixed the values of the groups. For instance, for two samples comparison, we gave the value of – 1 to one of the samples, and of 1 to the other samples. Then, we plotted our data, drew the squares and computed the ANOVA table. For regression, we also fixed the values in the horizontal axis, for instance, the doses of a drug or the number of exercise routines per day. All these are cases of *fixed effects*, as the effect we are trying to evaluate is fixed. The response (vertical axis) values are not fixed, and are usually normally distributed.

However, there are cases in which the effects are not fixed. For instance, we may want to test whether the amount of drug orally taken by mice affects their hours of sleep. We add the drug to their food but the amount of food eaten by each mouse will vary. Assuming that we can calculate how much drug each mouse took, we could build a regression line. However, the amount of drug taken is not fully controlled by us, as it has a random component. We call this a *random effect*. Random effect tests are a bit more complex as you have to deal with two levels of uncertainty, that of the independent and that of the dependent variable.

Last, if you extend the models we covered to multiple ways, these could be performed by considering some of these effects to be fixed, and some to be random. These are called *mixed effects* models, and can be very complex. This topic greatly exceeds our expectations, but since you are likely going to hear/read about mixed effect models, I wrote this section so you know what they are all about. Mixed effects models are one of the most active areas of research in statistical analysis.

Statistics Software

No one is expected to do all statistical tests by hand (although many of us have done quite a few). Thankfully, there are a number of statistical computing programs that facilitate the task. Probably you are mostly familiar with Microsoft Excel. Excel is not in itself a statistics software but a spreadsheet; however, it contains most of the statistical tests you may need. For some tests, you need to install the Analysis ToolPack. Other spreadsheets like Gnumeric and LibreOffice Calc also have similar features.

Other, more professional, programs are SPSS and Minitab, which allow you to do more sophisticated tests and have extra features like multiple testing, normality tests and more. The main disadvantage is that these are paywalled. The best option (probably) is R. R is a programming language specialized in statistical analysis. There are tons of books and resources (see next lesson) about R. It runs in the command-line (i.e. you have to write the instructions), yet it is quite intuitive, once you know the basics.

Before We Finish . . .

Let me summarize what we covered.

I hope you enjoyed this book as much as I enjoyed writing it. It is true that there were moments of frustration, in particular when I tried to explain complex tests and concepts without using a single formula. But I think I managed to do it after some work.

If you read the book and did the exercises, you should have a good understanding of how statistics work, you should know the main statistical concepts and should be able to perform the most common tests with small datasets. However, you should also know by now that you need larger sample sizes than those used in the examples in this book. The graphic approach used here to illustrate how statistics work is not the best approach when facing real and large datasets. For that, you need formulas (which as you surely suspect already will be full of *squared* terms) and some statistics software.

If you read this book out of curiosity, I hope you have a good idea now of how (and why) statistics works. If you are a student or professional, willing to learn more and start doing statistics as a pro, there are a few books and resources that may help you. I often recommend my students to read Dytham's *Choosing and Using Statistics*. It is a textbook aimed to undergraduate students and contains instructions to perform statistical tests in the most commonly used software. But there are other textbooks that cover basic statistics with examples, like McKullip's *Statistics Explained* or Alan Grafen and Rosie Hails' masterpiece *Modern Statistics for the Life Sciences*. The latter is out of print and a bit more advanced that the other books, but the treatment of statistics is analogous to the treatment in this book: statistical tests as linear models. The other classic book that influenced the way I wrote this book is Sokal and Rohlf's *Biometry*, and you should be able to find copies of it in almost any university library.

For computing *P* values and exploring various statistical tests, I strongly recommend that you start using R (www.r-project.org/), but for the first tests, you may explore Microsoft Excel and web sites like https://stattrek.com.

I believe my work here is done. Now it's your turn. Farewell.

Statistical Tables

We generally use computers to calculate the probability values associated with specific distributions and other quantities of statistical value. However, until recently, these values were often found in the so-called statistical tables. Given the nature of this book, I believe it is convenient for you to have some of these tables so you can do the exercises on the spot without the need for a computer or your phone. After all, I promised that you will only need graph paper, pencils and a ruler (and this book, of course).

Statistical Table 1. *F* Distribution Probabilities With 1 df in the Numerator

The following table gives the probability of rejecting the null hypothesis associated with an F distribution with one degree of freedom in the numerator (which will be most of our cases as linear models usually have one degree of freedom) for degrees of freedom from 2 to 12 in the denominator (columns), which will be determined by the sample size. The F values are in the first column, and the probability values are coloured for convenience (light-grey if $P < 0.05$, dark-grey if $P < 0.01$ and black if $P < 0.001$). For non-tabulated F values, you can estimate an interval, for instance, for an F value of 27 with one and three degrees of freedom, the P value lies within 0.012 and 0.015 (corresponding to the F values of 25 and 30).

F value	Degrees of freedom (denominator)										
	2	3	4	5	6	7	8	9	10	11	12
1	0.423	0.391	0.374	0.363	0.356	0.351	0.347	0.343	0.341	0.339	0.337
2	0.293	0.252	0.230	0.216	0.207	0.200	0.195	0.191	0.188	0.185	0.183
3	0.225	0.182	0.158	0.144	0.134	0.127	0.122	0.117	0.114	0.111	0.109
4	0.184	0.139	0.116	0.102	0.092	0.086	0.081	0.077	0.073	0.071	0.069
5	0.155	0.111	0.089	0.076	0.067	0.060	0.056	0.052	0.049	0.047	0.045
6	0.134	0.092	0.070	0.058	0.050	0.044	0.040	0.037	0.034	0.032	0.031
7	0.118	0.077	0.057	0.046	0.038	0.033	0.029	0.027	0.024	0.023	0.021
8	0.106	0.066	0.047	0.037	0.030	0.025	0.022	0.020	0.018	0.016	0.015
9	0.095	0.058	0.040	0.030	0.024	0.020	0.017	0.015	0.013	0.012	0.011
10	0.087	0.051	0.034	0.025	0.020	0.016	0.013	0.012	0.010	0.009	0.008
11	0.080	0.045	0.029	0.021	0.016	0.013	0.011	0.009	0.008	0.007	0.006
12	0.074	0.041	0.026	0.018	0.013	0.010	0.009	0.007	0.006	0.005	0.005
13	0.069	0.037	0.023	0.015	0.011	0.009	0.007	0.006	0.005	0.004	0.004
14	0.065	0.033	0.020	0.013	0.010	0.007	0.006	0.005	0.004	0.003	0.003
15	0.061	0.030	0.018	0.012	0.008	0.006	0.005	0.004	0.003	0.003	0.002
16	0.057	0.028	0.016	0.010	0.007	0.005	0.004	0.003	0.003	0.002	0.002
17	0.054	0.026	0.015	0.009	0.006	0.004	0.003	0.003	0.002	0.002	0.001
18	0.051	0.024	0.013	0.008	0.005	0.004	0.003	0.002	0.002	0.001	0.001
19	0.049	0.022	0.012	0.007	0.005	0.003	0.002	0.002	0.001	0.001	0.001
20	0.047	0.021	0.011	0.007	0.004	0.003	0.002	0.002	0.001	0.001	0.001
21	0.044	0.020	0.010	0.006	0.004	0.003	0.002	0.001	0.001	0.001	0.001
22	0.043	0.018	0.009	0.005	0.003	0.002	0.002	0.001	0.001	0.001	0.001
23	0.041	0.017	0.009	0.005	0.003	0.002	0.001	0.001	0.001	0.001	0.000
24	0.039	0.016	0.008	0.004	0.003	0.002	0.001	0.001	0.001	0.000	0.000
25	0.038	0.015	0.007	0.004	0.002	0.002	0.001	0.001	0.001	0.000	0.000
30	0.032	0.012	0.006	0.003	0.002	0.001	0.001	0.000	0.000	0.000	0.000
35	0.027	0.010	0.004	0.002	0.001	0.001	0.000	0.000	0.000	0.000	0.000
40	0.024	0.008	0.003	0.001	0.001	0.000	0.000	0.000	0.000	0.000	0.000
50	0.019	0.006	0.002	0.001	0.000	0.000	0.000	0.000	0.000	0.000	0.000
75	0.013	0.003	0.001	0.000	0.000	0.000	0.000	0.000	0.000	0.000	0.000
100	0.010	0.002	0.001	0.000	0.000	0.000	0.000	0.000	0.000	0.000	0.000

Statistical Table 2. Confidence Interval (95%) Estimations for Two-Sample Tests

For a given sum of squares (SS) and a specific sample size (n), you can find the quantity to be added and subtracted to the estimated mean to generate a confidence interval. For instance, for a mean of 5, that was estimated from four samples and that had an associated sum of squares of 12, we have to add/subtract 3.12 to the estimate. Therefore, the 95% confidence interval of the mean estimation will be [1.88–8.12].

Sum of Squares (SS)	Sample size (n) 3	4	5	6	7	8	9	10	11	12	13	14	15
0	0.00	0.00	0.00	0.00	0.00	0.00	0.00	0.00	0.00	0.00	0.00	0.00	0.00
1	1.76	0.92	0.62	0.47	0.38	0.32	0.27	0.24	0.21	0.19	0.17	0.16	0.15
2	2.48	1.30	0.88	0.66	0.53	0.45	0.38	0.34	0.30	0.27	0.25	0.23	0.21
3	3.04	1.59	1.08	0.81	0.65	0.55	0.47	0.41	0.37	0.33	0.30	0.28	0.26
4	3.61	1.84	1.24	0.94	0.76	0.63	0.54	0.48	0.42	0.38	0.35	0.32	0.30
5	3.93	2.05	1.39	1.05	0.84	0.71	0.61	0.53	0.48	0.43	0.39	0.36	0.33
6	4.30	2.25	1.52	1.15	0.92	0.77	0.67	0.58	0.52	0.47	0.43	0.39	0.36
7	4.65	2.43	1.64	1.24	1.00	0.84	0.72	0.63	0.56	0.51	0.46	0.42	0.39
8	4.97	2.60	1.76	1.33	1.07	0.89	0.77	0.67	0.60	0.54	0.49	0.45	0.42
9	5.27	2.76	1.86	1.41	1.13	0.95	0.82	0.72	0.64	0.57	0.52	0.48	0.44
10	5.55	2.91	1.96	1.48	1.19	1.00	0.86	0.75	0.67	0.61	0.55	0.51	0.47
11	5.83	3.05	2.06	1.56	1.25	1.05	0.90	0.79	0.70	0.64	0.58	0.53	0.49
12	6.08	3.18	2.15	1.63	1.31	1.09	0.94	0.83	0.74	0.66	0.60	0.55	0.51
13	6.33	3.31	2.24	1.69	1.36	1.14	0.98	0.86	0.77	0.69	0.63	0.58	0.53
14	6.57	3.44	2.32	1.76	1.41	1.18	1.02	0.89	0.79	0.72	0.65	0.60	0.55
15	6.80	3.56	2.40	1.82	1.46	1.22	1.05	0.92	0.82	0.74	0.68	0.62	0.57
16	7.03	3.67	2.48	1.88	1.51	1.26	1.09	0.95	0.85	0.77	0.70	0.64	0.59
17	7.24	3.79	2.56	1.94	1.56	1.30	1.12	0.98	0.88	0.79	0.72	0.66	0.61
18	7.45	3.90	2.63	1.99	1.60	1.34	1.15	1.01	0.90	0.81	0.74	0.68	0.63
19	7.66	4.00	2.71	2.05	1.65	1.38	1.18	1.04	0.93	0.84	0.76	0.70	0.65
20	7.86	4.11	2.78	2.10	1.69	1.41	1.22	1.07	0.95	0.86	0.78	0.72	0.66
21	8.05	4.21	2.85	2.15	1.73	1.45	1.25	1.09	0.97	0.88	0.80	0.73	0.68
22	8.24	4.31	2.91	2.20	1.77	1.48	1.27	1.12	1.00	0.90	0.82	0.75	0.69
23	8.42	4.41	2.98	2.25	1.81	1.52	1.30	1.14	1.02	0.92	0.84	0.77	0.71
24	8.61	4.50	3.04	2.30	1.85	1.55	1.33	1.17	1.04	0.94	0.85	0.78	0.73
25	8.78	4.59	3.10	2.35	1.89	1.58	1.36	1.19	1.06	0.96	0.87	0.80	0.74
26	8.96	4.68	3.17	2.39	1.93	1.61	1.39	1.22	1.08	0.98	0.89	0.82	0.75
27	9.13	4.77	3.23	2.44	1.96	1.64	1.41	1.24	1.10	1.00	0.91	0.83	0.77
28	9.29	4.86	3.29	2.48	2.00	1.67	1.44	1.26	1.12	1.01	0.92	0.85	0.78
29	9.46	4.95	3.34	2.53	2.03	1.70	1.46	1.28	1.14	1.03	0.94	0.86	0.80
30	9.62	5.03	3.40	2.57	2.07	1.73	1.49	1.31	1.16	1.05	0.96	0.88	0.81
31	9.78	5.12	3.46	2.61	2.10	1.76	1.51	1.33	1.18	1.07	0.97	0.89	0.82
32	9.94	5.20	3.51	2.65	2.14	1.79	1.54	1.35	1.20	1.08	0.99	0.91	0.84
33	10.09	5.28	3.57	2.70	2.17	1.82	1.56	1.37	1.22	1.10	1.00	0.92	0.85
34	10.24	5.36	3.62	2.74	2.20	1.84	1.58	1.39	1.24	1.12	1.02	0.93	0.86
35	10.39	5.44	3.67	2.78	2.23	1.87	1.61	1.41	1.26	1.13	1.03	0.95	0.88
36	10.54	5.51	3.72	2.82	2.27	1.90	1.63	1.43	1.27	1.15	1.05	0.96	0.89
37	10.68	5.59	3.78	2.85	2.30	1.92	1.65	1.45	1.29	1.17	1.06	0.97	0.90
38	10.83	5.66	3.83	2.89	2.33	1.95	1.68	1.47	1.31	1.18	1.08	0.99	0.91
39	10.97	5.74	3.88	2.93	2.36	1.97	1.70	1.49	1.33	1.20	1.09	1.00	0.92
40	11.11	5.81	3.93	2.97	2.39	2.00	1.72	1.51	1.34	1.21	1.10	1.01	0.94
50	12.42	6.50	4.39	3.32	2.67	2.23	1.92	1.69	1.50	1.35	1.23	1.13	1.05
60	13.61	7.12	4.81	3.64	2.92	2.45	2.11	1.85	1.65	1.48	1.35	1.24	1.15
70	14.70	7.69	5.19	3.93	3.16	2.64	2.27	2.00	1.78	1.60	1.46	1.34	1.24
80	15.71	8.22	5.55	4.20	3.38	2.83	2.43	2.13	1.90	1.71	1.56	1.43	1.32
90	16.66	8.72	5.89	4.45	3.58	3.00	2.58	2.26	2.02	1.82	1.65	1.52	1.40
100	17.57	9.19	6.21	4.69	3.78	3.16	2.72	2.38	2.12	1.92	1.74	1.60	1.48

Statistical Table 3. *F* Distribution Probabilities with 2 df in the Numerator

Like Statistical Table 1, this table indicates the *P* values associated to an *F* distribution, but in this case with two degrees of freedom in the numerator (see text in Statistical Table 1).

F value	Degrees of freedom (denominator)										
	2	3	4	5	6	7	8	9	10	11	12
1	0.500	0.465	0.444	0.431	0.422	0.415	0.410	0.405	0.402	0.399	0.397
2	0.333	0.281	0.250	0.230	0.216	0.206	0.198	0.191	0.186	0.182	0.178
3	0.250	0.192	0.160	0.139	0.125	0.115	0.107	0.100	0.095	0.091	0.088
4	0.200	0.142	0.111	0.092	0.079	0.069	0.063	0.057	0.053	0.049	0.047
5	0.167	0.111	0.082	0.064	0.053	0.045	0.039	0.035	0.031	0.029	0.026
6	0.143	0.089	0.063	0.047	0.037	0.030	0.026	0.022	0.019	0.017	0.016
7	0.125	0.074	0.049	0.036	0.027	0.021	0.017	0.015	0.013	0.011	0.010
8	0.111	0.063	0.040	0.028	0.020	0.016	0.012	0.010	0.008	0.007	0.006
9	0.100	0.054	0.033	0.022	0.016	0.012	0.009	0.007	0.006	0.005	0.004
10	0.091	0.047	0.028	0.018	0.012	0.009	0.007	0.005	0.004	0.003	0.003
11	0.083	0.042	0.024	0.015	0.010	0.007	0.005	0.004	0.003	0.002	0.002
12	0.077	0.037	0.020	0.012	0.008	0.005	0.004	0.003	0.002	0.002	0.001
13	0.071	0.033	0.018	0.010	0.007	0.004	0.003	0.002	0.002	0.001	0.001
14	0.067	0.030	0.016	0.009	0.005	0.004	0.002	0.002	0.001	0.001	0.001
15	0.063	0.027	0.014	0.008	0.006	0.003	0.002	0.001	0.001	0.001	0.001
16	0.059	0.025	0.012	0.007	0.004	0.002	0.002	0.001	0.001	0.001	0.000
17	0.056	0.023	0.011	0.006	0.003	0.002	0.001	0.001	0.001	0.000	0.000
18	0.053	0.021	0.010	0.005	0.003	0.002	0.001	0.001	0.000	0.000	0.000
19	0.050	0.020	0.009	0.005	0.003	0.001	0.001	0.001	0.000	0.000	0.000
20	0.048	0.018	0.008	0.004	0.002	0.001	0.001	0.000	0.000	0.000	0.000
21	0.045	0.017	0.008	0.004	0.002	0.001	0.001	0.000	0.000	0.000	0.000
22	0.043	0.016	0.007	0.003	0.002	0.001	0.001	0.000	0.000	0.000	0.000
23	0.042	0.015	0.006	0.003	0.002	0.001	0.000	0.000	0.000	0.000	0.000
24	0.040	0.014	0.006	0.003	0.001	0.001	0.000	0.000	0.000	0.000	0.000
25	0.038	0.013	0.005	0.002	0.001	0.001	0.000	0.000	0.000	0.000	0.000
30	0.032	0.010	0.004	0.002	0.001	0.000	0.000	0.000	0.000	0.000	0.000
35	0.028	0.008	0.003	0.001	0.000	0.000	0.000	0.000	0.000	0.000	0.000
40	0.024	0.007	0.002	0.001	0.000	0.000	0.000	0.000	0.000	0.000	0.000
50	0.020	0.005	0.001	0.000	0.000	0.000	0.000	0.000	0.000	0.000	0.000
75	0.013	0.003	0.001	0.000	0.000	0.000	0.000	0.000	0.000	0.000	0.000
100	0.010	0.002	0.000	0.000	0.000	0.000	0.000	0.000	0.000	0.000	0.000

Statistical Table 4. *F* Distribution Probabilities with 3 df in the Numerator

Like Statistical Table 1, this table indicates the *P* values associated to an *F* distribution, but in this case with three degrees of freedom in the numerator (see text in Statistical Table 1).

F value	Degrees of freedom (denominator)										
	2	3	4	5	6	7	8	9	10	11	12
1	0.535	0.500	0.479	0.465	0.455	0.447	0.441	0.436	0.432	0.429	0.426
2	0.350	0.292	0.250	0.233	0.216	0.203	0.193	0.185	0.178	0.172	0.168
3	0.260	0.196	0.158	0.134	0.117	0.105	0.095	0.088	0.082	0.077	0.073
4	0.206	0.142	0.107	0.085	0.070	0.060	0.052	0.046	0.041	0.038	0.035
5	0.171	0.110	0.077	0.050	0.045	0.037	0.031	0.026	0.023	0.020	0.018
6	0.148	0.088	0.058	0.041	0.031	0.024	0.019	0.016	0.013	0.011	0.010
7	0.128	0.072	0.045	0.031	0.022	0.016	0.013	0.010	0.009	0.007	0.006
8	0.113	0.061	0.038	0.024	0.016	0.012	0.009	0.007	0.005	0.004	0.003
9	0.102	0.052	0.030	0.019	0.012	0.008	0.006	0.005	0.003	0.003	0.002
10	0.092	0.045	0.025	0.015	0.009	0.006	0.004	0.003	0.002	0.002	0.001
11	0.084	0.040	0.021	0.012	0.007	0.005	0.003	0.002	0.002	0.001	0.001
12	0.078	0.035	0.018	0.010	0.006	0.004	0.002	0.002	0.001	0.001	0.001
13	0.072	0.032	0.016	0.008	0.005	0.003	0.002	0.001	0.001	0.001	0.000
14	0.067	0.029	0.014	0.007	0.004	0.002	0.002	0.001	0.001	0.000	0.000
15	0.062	0.026	0.012	0.006	0.003	0.002	0.001	0.001	0.000	0.000	0.000
16	0.059	0.024	0.011	0.005	0.003	0.002	0.001	0.001	0.000	0.000	0.000
17	0.056	0.022	0.010	0.005	0.002	0.001	0.001	0.000	0.000	0.000	0.000
18	0.053	0.020	0.009	0.004	0.002	0.001	0.001	0.000	0.000	0.000	0.000
19	0.050	0.019	0.008	0.004	0.002	0.001	0.001	0.000	0.000	0.000	0.000
20	0.048	0.017	0.007	0.003	0.002	0.001	0.000	0.000	0.000	0.000	0.000
21	0.046	0.016	0.007	0.003	0.001	0.001	0.000	0.000	0.000	0.000	0.000
22	0.044	0.015	0.006	0.003	0.001	0.001	0.000	0.000	0.000	0.000	0.000
23	0.042	0.014	0.006	0.002	0.001	0.001	0.000	0.000	0.000	0.000	0.000
24	0.040	0.013	0.005	0.002	0.001	0.000	0.000	0.000	0.000	0.000	0.000
25	0.039	0.013	0.005	0.002	0.001	0.000	0.000	0.000	0.000	0.000	0.000
30	0.032	0.010	0.003	0.001	0.001	0.000	0.000	0.000	0.000	0.000	0.000
35	0.028	0.008	0.002	0.001	0.000	0.000	0.000	0.000	0.000	0.000	0.000
40	0.024	0.006	0.002	0.001	0.000	0.000	0.000	0.000	0.000	0.000	0.000
50	0.020	0.005	0.001	0.000	0.000	0.000	0.000	0.000	0.000	0.000	0.000
75	0.013	0.003	0.001	0.000	0.000	0.000	0.000	0.000	0.000	0.000	0.000
100	0.010	0.002	0.000	0.000	0.000	0.000	0.000	0.000	0.000	0.000	0.000

Statistical Table 5. Chi-Squared Distribution Probabilities

The following table gives the probability of rejecting the null hypothesis associated with a chi-squared distribution for degrees of freedom from 1 to 100. The X^2 values are in the first column, and the probability values are coloured for convenience (light-grey if $P < 0.05$, dark-grey if $P < 0.01$ and black if $P < 0.001$). Like in Statistical Table 1, for non-tabulated F values, you can estimate an interval (see text in Statistical Table 1).

X^2 value	Degrees of freedom										
	1	2	3	4	5	6	7	8	9	10	11
1	0.317	0.607	0.801	0.910	0.963	0.986	0.995	0.998	0.999	1.000	1.000
2	0.157	0.368	0.572	0.736	0.849	0.920	0.960	0.981	0.991	0.996	0.998
3	0.083	0.223	0.392	0.558	0.700	0.809	0.885	0.934	0.964	0.981	0.991
4	0.046	0.135	0.261	0.406	0.549	0.677	0.780	0.857	0.911	0.947	0.970
5	0.025	0.082	0.172	0.287	0.416	0.544	0.660	0.758	0.834	0.891	0.931
6	0.014	0.050	0.112	0.199	0.306	0.423	0.540	0.647	0.740	0.815	0.873
7	0.008	0.030	0.072	0.136	0.221	0.321	0.429	0.537	0.637	0.725	0.799
8	0.005	0.018	0.046	0.092	0.156	0.238	0.333	0.433	0.534	0.629	0.713
9	0.003	0.011	0.029	0.061	0.109	0.174	0.253	0.342	0.437	0.532	0.622
10	0.002	0.007	0.019	0.040	0.075	0.125	0.189	0.265	0.350	0.440	0.530
11	0.001	0.004	0.012	0.027	0.051	0.088	0.139	0.202	0.276	0.358	0.443
12	0.001	0.002	0.007	0.017	0.035	0.062	0.101	0.151	0.213	0.285	0.364
13	0.000	0.002	0.005	0.011	0.023	0.043	0.072	0.112	0.163	0.224	0.293
14	0.000	0.001	0.003	0.007	0.016	0.030	0.051	0.082	0.122	0.173	0.233
15	0.000	0.001	0.002	0.005	0.010	0.020	0.036	0.059	0.091	0.132	0.182
16	0.000	0.000	0.001	0.003	0.007	0.014	0.025	0.042	0.067	0.100	0.141
17	0.000	0.000	0.001	0.002	0.004	0.009	0.017	0.030	0.049	0.074	0.108
18	0.000	0.000	0.000	0.001	0.003	0.006	0.012	0.021	0.035	0.055	0.082
19	0.000	0.000	0.000	0.001	0.002	0.004	0.008	0.015	0.025	0.040	0.061
20	0.000	0.000	0.000	0.000	0.001	0.003	0.006	0.010	0.018	0.029	0.045
21	0.000	0.000	0.000	0.000	0.001	0.002	0.004	0.007	0.013	0.021	0.033
22	0.000	0.000	0.000	0.000	0.001	0.001	0.003	0.005	0.009	0.015	0.024
23	0.000	0.000	0.000	0.000	0.000	0.001	0.002	0.003	0.006	0.011	0.018
24	0.000	0.000	0.000	0.000	0.000	0.001	0.001	0.002	0.004	0.008	0.013
25	0.000	0.000	0.000	0.000	0.000	0.000	0.001	0.002	0.003	0.005	0.009
30	0.000	0.000	0.000	0.000	0.000	0.000	0.000	0.000	0.000	0.001	0.002
35	0.000	0.000	0.000	0.000	0.000	0.000	0.000	0.000	0.000	0.000	0.000
40	0.000	0.000	0.000	0.000	0.000	0.000	0.000	0.000	0.000	0.000	0.000
50	0.000	0.000	0.000	0.000	0.000	0.000	0.000	0.000	0.000	0.000	0.000
75	0.000	0.000	0.000	0.000	0.000	0.000	0.000	0.000	0.000	0.000	0.000
100	0.000	0.000	0.000	0.000	0.000	0.000	0.000	0.000	0.000	0.000	0.000

Solution to Exercises

Lesson 1. Build histograms and box plots for Alice's and Bob's datasets:

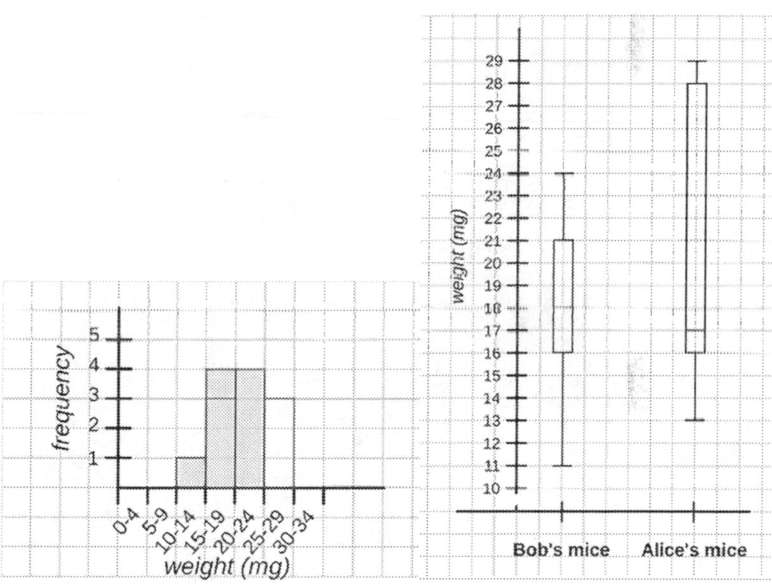

Lesson 2. Bob's mice mean and variance are 18 and 16, respectively. Alice's mice mean and variance are 21 and 42, respectively.

Lesson 3. Bob's estimates of mice population mean and variance are 18 and 18, respectively. Alice's estimates of mice population mean and variance are 21 and 49, respectively.

Lesson 4. The null hypothesis is that Alice's and Bob's mice weights are not different. In other (more technical) words, Alice's and Bob's samples come from the same population of mice. The alternative hypothesis is that the two samples have significantly different weights. In other (again, more technical) words, Alice's and Bob's samples come from different populations (each one with its own, different, average weight).

Lesson 5. ANOVA table for Bob's mice:

	SS	df	MS	F	*P*
MODEL	81	1	81	3.9	0.086–0.127
ERROR	144	7	20.6		
TOTAL	225	8			

The range of *P* values comes from Statistical Table 1. The exact *P* value is 0.089.

Lesson 6. ANOVA table for Bob's one-sample experiment:

	SS	df	MS	F	*P*
MODEL	405	1	405	60.8	0.003–0.006
ERROR	20	3	6.7		
TOTAL	425	4			

The range of *P* values comes from Statistical Table 1. The exact *P* value is 0.004.

Lesson 7. ANOVA table for Bob's two-sample experiment:

	SS	df	MS	F	*P*
MODEL	160	1	160	32	0.002–0.001
ERROR	30	6	5		
TOTAL	190				

The range of *P* values comes from Statistical Table 1, which for a one-tailed test is half the value: 0.001–0.0005. The exact *P* value is 0.0005. The 95% confidence intervals for the estimate of the means of mice in high-fat and low-fat diets are, respectively, as computed using Statistical Table 2, [24.22–29.78] and [17.04–20.96] grams.

Lesson 8. As in the previous lesson exercise, the partition of the sums of squares is $160+30=190$.

Lesson 9. ANOVA table for Alice's suggested multiple-sample test is:

	SS	df	MS	F	*P*
MODEL	15	2	7.5	0.3	Over 0.415
ERROR	175	7	25		
TOTAL	190				

The exact *P* value is 0.750.

Lesson 10. ANOVA table of the two-way test comparing strain and diet:

	SS	df	MS	F	*P*
MODEL (strain)	1	2	0.5	0.10	Over 0.397
MODEL (diet)	162	1	162	33.52	0.000
INTERACTION	7	2	3.5	0.72	Over 0.397
ERROR	58	12	4.8		
TOTAL	228	17			

It can be concluded that, statistically, only the diet has a detectable effect on the mice weight.

Lesson 11. ANOVA table for Alice's regression test is:

	SS	df	MS	F	*P*
MODEL	102	1	102	102	0.000
ERROR	8	8	1		
TOTAL	110	9			

Which indicates, pretty clearly, that the more the time exercising, the lower the mice weight.

Lesson 12. ANOVA table for Bob's non-parametric two samples test is:

	SS	df	MS	F	*P*
MODEL	62.5	1	62.5	25	0.002–0.001
ERROR	20.0	8	2.5		
TOTAL	82.5				

The exact *P* value is 0.001.

Lesson 13. The categorical test table for Alice's mice food preference is:

Supermarket	Observed	Expected	Squared Difference	Standardized Squared Difference
Trisco	20	10	100	10
Aldo	5	10	25	2.5
Wait-flower	5	10	25	2.5

The X^2 value is 15 which, for two degrees of freedom, give us a *P* value of 0.001 (Statistical Table 5). We have statistical support that the mice have a preference when choosing their food.

Index

Note: Page numbers in *italics* refer to a figure on the corresponding page.

Printed in the United States
by Baker & Taylor Publisher Services